Opportunities for Innovation

POLLUTION PREVENTION

Edited by:

David E. Edgerly
National Institute of Standards and Technology
Gaithersburg, MD 20899

Prepared for:

U.S. Department of Commerce
National Institute of Standards and Technology
Gaithersburg, MD 20899

LANCASTER · BASEL

TECHNOMIC Publishing Company, Inc.
851 New Holland Avenue, Box 3535, Lancaster, Pennsylvania 17604, USA

TECHNOMIC Publishing AG
Missionsstrasse 44, CH-4055 Basel, Switzerland

ISBN 1-56676-287-1

CONTENTS

	PAGE
Preface... Steven T. Ostheim, Pittsburgh, PA	v
The Challenges & Opportunities for Pollution Prevention.. Robert B. Pojasek, Winchester, MA	1
Pollution Prevention in the Chemical Industry............ Thomas W. Zosel, St. Paul, MN	13
Pollution Prevention in the Electronic and Office Equipment Industry... Patricia A. Calkins, Jack C. Azar, Webster, NY	26
Pollution Prevention in the Metals Coating Industry...... Marvin M. Floer, Auburn Hills, MI	45
Pollution Prevention in the Metal Degreasing Industry.... Stephen Evanoff, Ft. Worth, TX	56
Pollution Prevention in the Metal Finishing Industry..... Kevin P. Vidmar, East Greenwich, RI	75
Pollution Prevention in the Pulp and Paper Industry...... David H. Critchfield, Jay, ME	107
Pollution Prevention in the Printing Industry............ C. Nelson Ho, Pittsburgh, PA	118
Pollution Prevention in the Textile Industry............. Brent Smith, Raleigh, NC	131

PREFACE

The objective of NIST's Opportunities for Innovative (OFI) is to encourage the industrial competitiveness of small- to medium-sized businesses. U.S. businesses, small as well as large, face continuing competitive and financial challenges, including the growing cost to comply with increasing environmental regulations, including the costs associated with the generation and disposal of wastes. Pollution prevention (P2), reducing pollution at the source, is a sound way for businesses to reduce their environmental compliance costs, while simultaneously helping the environment and increasing their production efficiencies (by lowering their waste generation rates and inefficient use of energy resources).

The objective in preparing this monograph was to identify technological opportunities within a number of selected industries and/or manufacturing/finishing processes, to reduce pollution. These industries/processes were selected as representative of and applicable to the broad range of U.S. manufacturing businesses. These include: metals coating (i.e., painting) which is widely done in a number of major industries, including the automotive industry; metals degreasing; recyclability of office equipment; chemical manufacturing; printing; textiles dye and dyeing; and the pulp and paper industry. Solutions developed for these industries should be broadly applicable across a number of industrial sectors. Additionally, the promulgation of new regulations, requiring companies to change their historical manufacturing and waste generating/disposal practices, will continue to create technological needs and opportunities for savvy businesses to develop solutions.

To identify these technology opportunities and solutions, experts within these industries were chosen to author various chapters. They were asked to identify problems and needs, and possible areas of solutions, if possible. We believe the authors have been successful in identifying a number of technology development and/or process modification needs which could represent collaborative opportunities between large and smaller industry and solution providers.

In completing this monograph, I am indebted to Paula Comella and Mary Rose Glaser for their technical editing expertise, and especially grateful to Audra Ometz, for the many hours she spent typing the draft and final edited chapters.

Finally, in addition to the authors and reviewers, Jarda Ulbrecht from the National Institute of Standards and Technology played a key role in the preparation of this Monograph. This Monograph is a testament to his vision for the OFI program, and for recognizing the integral role of environmental issues in the competitiveness of U.S. industry. We are honored to have been able to participate in this Monograph.

Steven T. Ostheim
Volume Editor and Director
Center for Hazardous Materials Research
Subsidiary of the University of Pittsburgh Trust
Pittsburgh, PA

The Challenges and Opportunities of Pollution Prevention
Robert B. Pojasek, Ph.D.
Corporate Vice President
GEI Consultants, Inc.
Winchester, Massachusetts 01890-1970
(617) 721-4097

Pollution prevention is not just about producing less waste; it involves conserving the use of hazardous materials and other resources such as air, water, and energy. It also involves the most efficient use of all of these resources in the manufacturing process. To properly audit a manufacturing process to identify pollution prevention opportunities, a team should be assembled to prepare detailed process flow documentation of the facilities' operations and processes (the Descriptive Approach). While there are impediments to implementing pollution prevention recommendations, facilities should accept and learn to operate with the impediments. The best way to overcome these impediments is to foster the notion of continuous improvement in manufacturing firms through ongoing programs rather than discrete technological solutions. Government can support pollution prevention by encouraging firms to use the "Descriptive Method" which encourages firms to understand process functionality, rather than its prescriptive checklists, etc.

Key Words

Benchmarking; biotechnology-based processes; coating removal; descriptive approach; impediments to pollution prevention; improved materials; process monitoring systems; separations technology; surface preparation; surfact treatment; technology avoidance.

1. Introduction

Unfortunately, changing a manufacturing process to accommodate a new technology that encourages pollution prevention is never an easy decision for a company to make. Built-in resistance to change, as typified by the "not invented here syndrome," is difficult to overcome, even in facilities with pollution prevention programs that serve as "best in class" examples for others to benchmark against. However, many firms simply fail to either realize the opportunities for pollution prevention which potentially include new technologies or to recognize the ability of these "clean" technologies to provide a reasonable return on investment taking into account activity-based costs.

There is a growing awareness of areas where the potential for invention would greatly improve pollution prevention efforts. R&D-based firms involved in meeting these technological needs must understand how to best present their equipment or materials. They must keep in mind both the cultural element of how the facility operates and the proper means of justifying the costs, taking environmental activity-based costs typically allocated to the facilities overhead structure. Realizing savings will help lower the facility's overhead rate. This will be looked upon in a favorable light by any facility manager. Hopefully, by understanding the concepts presented in this chapter, no matter how sobering they may be, development and acceptance of new pollution prevention technologies can help advance the competitiveness of both parties in a global market.

2. Current Perspectives on Pollution Prevention

Many practioners perceive pollution prevention as a well-defined practice complete with its own literature of case histories and its protocols of checklists, questionnaires, and worksheets. The U.S. Environmental Protection Agency (EPA) and a number of states have sought to standardize the practice within their jurisdictions by publishing guides for companies to follow. Such a prescriptive approach to pollution prevention often leads to opportunistic fixes. Companies are encouraged to look in places where others have looked before and only try solutions that others have written about. Emulation of case histories leads to an overreliance on good operating practices and, to some extent, materials substitution. This does not create a demand for new technologies.

Another problem with the way pollution prevention is practiced is that it has been conducted largely by the environmental personnel. Pollution prevention is not just about producing less waste; it involves the conservation of the use of hazardous materials and other resources such as air, water, and energy. It also involves the most efficient use of all these resources in the manufacturing process. These are not areas where the environmental specialists have a lot of experience. Only through a thorough knowledge of the process can a cross-functional team catalog all losses of resources from the process. By using the principles of continuous improvement, the firm can sequentially address the opportunities while moving towards the use of new technologies. Process vendors will want to exploit efficiency. A strong theme in most manufacturer's drive is to remain competitive in a global economy.

Pollution prevention literature does not adequately address the function of a manufacturing process. Each operation or process in industry consists of a functional sequence of events. One action initiates another, which in turn initiates still another, until the process is completed with some type of product or result. By understanding functionality, both the company and the technology vendor can better understand the need for new technologies in the process. The "Descriptive Approach" to pollution prevention [1] is specifically designed to meet this need. This approach is briefly presented below.

To gain a proper appreciation of functionality, each firm should prepare detailed process flow documentation of all its operations and processes. This is the starting point of most quality improvement programs. It is best to assemble a team for this task, which includes the senior managers of the different functional groups at the facility (i.e., production, scheduling, purchasing, accounting, environmental, health and safety, quality improvement, labor, maintenance, facilities engineering, etc.) Using a trained facilitator to move the process along, this team will prepare preliminary process flow diagrams. Without an unbiased third party to guide assembly of the process flow documentation, the different perceptions individual team members have of the work process can prevent the agreement necessary to construct the diagram. The group will note the following items for each unit: the materials used, other resources used, activity-based costs, process losses, and the functionality of each unit.

It is important to identify the boundaries of the process. This is often easier said than done because the work often begins earlier than people realize and ends later. Easily overlooked are preparatory stages at the start and service, repair and other steps after work is thought to be complete. The group must be on the lookout for intermittent and ancillary operations.

This group should work on one segment of the process flow diagram at a time. When the overall process has been depicted, they must go back and provide the detail that may be involved in some of the steps. This detail is necessary to properly depict functionality. A storyboard format is good to use for this exercise. It can be left in place for a period of time to allow the group to observe the operations and then return to make changes or additions. Each of the team members will see the process differently. This is necessary to get a complete picture that none of the team members can provide alone.

When the initial exercise is complete, the process flow diagram will be cleaned up and reviewed by many others in the organization. They will likely offer worthwhile additions and corrections. The most important output from this activity is the list of losses from the operation. As shown in Fig. 1, these losses must include those to the air, water, solid wastes, as well as spills/leaks and accidents. These losses will be cataloged by the individual units in the process responsible for generating them.

Every loss from the operation is an opportunity for pollution prevention. It becomes readily apparent that every manufacturing operation has many process losses--perhaps more than are worth considering in a pollution prevention program. Pareto analysis can be used to rank these losses. This analysis demonstrates that 80 percent of the environmental activity-based costs are caused by 20 percent of the opportunities. Hence, the 80/20 rule.

Just as quality improvement teams operate in a total quality management program, separate teams can take on each of the primary opportunities. Groups dedicated to primary opportunities must be trained in group dynamics and creative thought to be successful in identifying all the alternatives for addressing losses. Pollution prevention programs show that there are many categories that can be considered (see Fig. 2). Some alternatives should be examined from each such category. A prize should be given to the group member who comes up with the "most outrageous idea that might just work" as voted by the other members in the group. The group will then look at a set of criteria to screen the large number of alternatives to a smaller number for the feasibility study.

It might be that one of the alternatives is an easy-to-implement operating practice. The group can recommend that this be done and observe the process before and after the change. If the loss was not eliminated or shifted to other media, it must be added back to the list of opportunities kept by the pollution prevention steering committee along with the information collected to date. In this manner, there will always be new opportunities to consider. Continuous improvement will be noticed as the progress proceeds to materials substitution and technology implementation. Technology will be used both in the recycle and pollution prevention modes.

It is important to say a word about the feasibility study at this point. In the world of pollution control, a feasibility study has traditionally meant choosing the best available control technology and implementing it. There was little choice because of short time frames and the

difficulty of getting regulatory agencies to approve alternative technologies. Facilities have gotten a little "rusty" when it comes to conducting a valid feasibility study [2]. For pollution prevention, all the alternatives for each opportunity must be considered. Included in this range of alternatives are the operating practices, materials substitution, product changes, technology changes, and recycle/reuse. If an improved operating practice can be implemented in the name of continuous improvement with little impact on the operation, it should be done in an expedient
fashion. However, new technologies may require bench- and pilot-testing and more detailed analysis before they can be utilized.

It is easy to see that attention to operating practices in the early implementation of pollution prevention is the first order of business. Often, significant reduction in the amount of waste generated will be realized by firms that use chemicals. Cleaning of batch processes and coating of materials are major consumers of chemicals. Technologies which make these operations more efficient or allow the recycling and reuse of spent materials should be in great demand.

Many companies seek substitute chemicals which do not appear on any government lists. Use of regulated materials adds substantial costs to an operation. However, when many firms switch to the same substitute, there may be unintended consequences [3]. Allergies and reproductive effects have played havoc with some commonly used substitutes. There may be a need for technology that allows the use of a chemical in a closed system. This allows much less exposure to the chemical and optimizes its recycling. Technologies which can clean surfaces without chemicals (e.g., laser, electron beam, CO_2 pellets, etc.) should also see increased demand.

It is easy to understand how the Descriptive Approach helps both the company and the technology developer appreciate the need and the benefit of a given technological alternative. Of course, the greater the level of understanding, the better the chance to make a match. It should be noted that this approach is not industry-specific. Opportunities to service many diverse industries with a similar problem can be determined.

Before discussing some general areas where matches may currently exist, it is important to examine the historic trend to avoid new technologies followed by the current regulatory impediments to the use of pollution prevention technology.

3. Historic Technology Avoidance

During the industrial revolution in the middle of the last century, industry produced a series of significant innovations in process and product technologies. In the early part of this century, manufacturers increasingly refined proven technology rather than developing and implementing new and diverse technologies to accomplish, or even eliminate, traditional tasks. This apparent trend toward a more stable, conservative approach to process technology in a broad range of industries caused a shift towards more modest improvements in productivity. There were some notable exceptions to this trend, including electronics, chemicals, and biotechnology. In these industries, most of the product breakthroughs depend on breakthroughs in process capabilities. However, in most other manufacturing industries, there was a demand on short-term profits rather than long-term development of product and process technologies [4].

Because of the feeling that new technology did not seem to offer great potential, manufacturing focused instead on product engineering at the expense of process engineering. Since manufacturers had their hands full simply adding capacity of a known type, they saw no pressing need to add new process technologies at the same time. Consequently, many firms

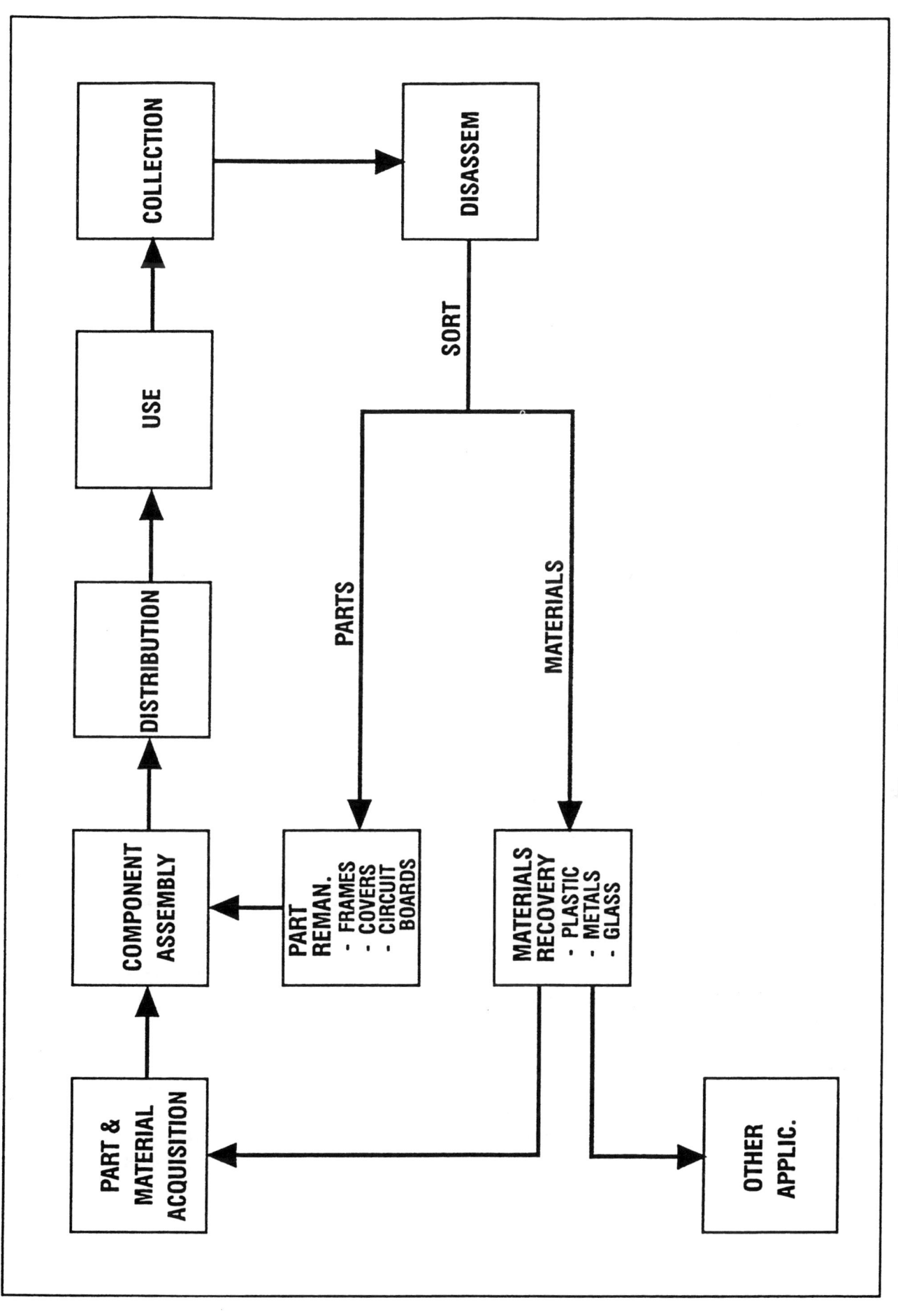

Figure 1. Closed-Loop Vision

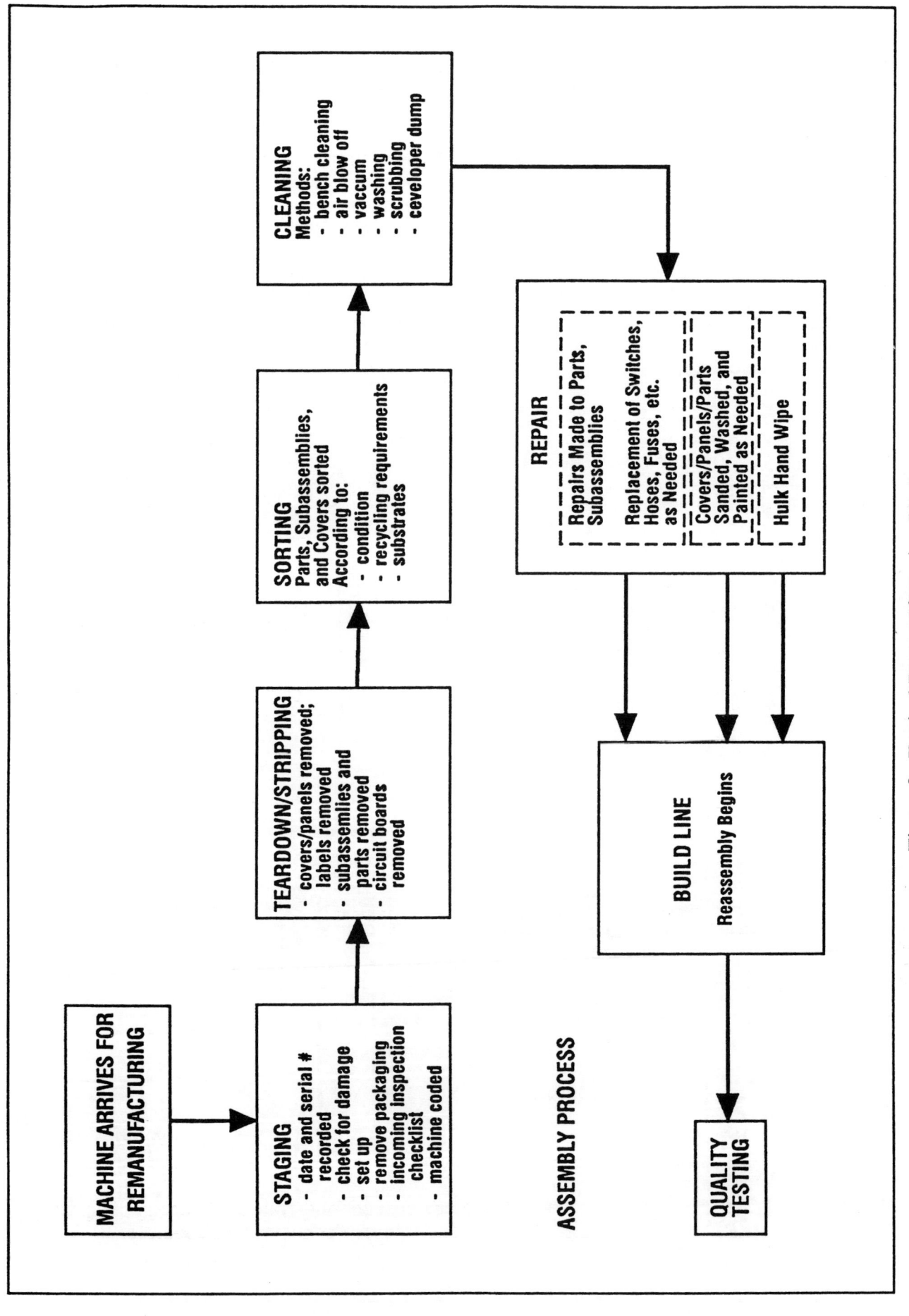

Figure 2. Typical Remanufacturing Process

spent incremental dollars on product technology and very little on new process technology. Process development was left to the equipment suppliers. Most firms allowed their own skills at such development to decline. The operative word was, "Make the product with no surprises."

By operating processes more efficiently, manufacturing managers increased productivity. This mode of operation is dramatically upset when significant improvement or new technology is implemented by a competitor. Management's view of manufacturing was based on the premise that smart people should be able to determine the optimal solution for handling the tasks of the manufacturing function and then control the process and organization for maximum stability and efficiency until some external event forces change. This is a reactive view that overlooks the potential contributions of the manufacturing function to overall competitiveness. Advances in production planning, project evaluation, and operations research offered new tools for maintaining stability and increasing productivity. Computers were able to take detailed measurements and exhibit sophisticated control. They became another impediment to process changes.

These manufacturing trends have lead to increased tuning and refining of a set of resources that were, in many cases, outdated and increasingly inappropriate. Some firms slipped into a debilitating spiral: additional investment was withheld because the current investment was not performing as expected; those operating the current investment simply tried to minimize the problem in the near term rather than looking for long-term solutions they knew would not be approved and supported. These conditions still linger in today's manufacturing industry.

It is interesting to note the attitudes towards technology in a benchmarking study of the facilities with the "best-in-class" pollution prevention performance. In a survey conducted by the Business Roundtable [5] of companies with pollution prevention programs, eight questions were asked regarding the use of new technologies.

> How do you find and evaluate new technology or pollution prevention ideas?
> What are your sources of information for new technology?
> (Internal technology transfer, external technology transfer, etc.)
>
> Please describe your technology transfer process (facility to facility transfer, corporate to facility transfer, external to the corporation [university, industry], etc.)
> What unique or innovative approach or technology have you used to produce significant results?
> How do you evaluate new technologies and lessons learned across processes?
> How are new technologies and lessons learned incorporated into your program?
> Are the gains you've experienced incremental or step changes or a combination of both? Please explain.
> What could be done to enhance technology transfer in the future?
>> Six benchmarked firms responded to these questions on the use of technology to achieve significant improvement as follows:

3M - Sources of new technology included corporate research group, other 3M plants, in-house development, other 3M divisions, and outside vendors and consultants.
DuPont - Sources of new technology included corporate engineering, company R&D, Texas Chemical Council, and company communication sources.

Intel - Sources of new technology were primarily internal R&D and process engineers and collaborative efforts with vendors.

Martin Marietta - Sources of new technology included seminars, trade journals, professional associations, and internal technology transfer meetings.

Monsanto - Sources of new technology included directed university research, intra-company technology transfer, and alliances with other companies.

Procter & Gamble - Sources of new technology came primarily from facility employees. Other sources included other company facilities, suppliers, and waste brokers.

None of these firms emphasized working with firms included in developing new technologies to help with their pollution prevention programs. There may be a number of regulatory impediments that might explain this finding.

4. Impediments to Pollution Prevention

The benefits of implementing pollution prevention have been widely promoted by EPA and the state governments. However, several impediments help create business risks for those who would seek to implement pollution prevention programs. It is beyond the scope of this chapter to examine all of the impediments because they vary greatly from industry to industry and from facility to facility within the same industry. However, it is worthwhile to explore some of the regulatory and nonregulatory impediments to see what impact they may have on the use of new pollution prevention technologies in industry.

It is important to point out that companies learn to operate with the impediments. In some cases, impediments may even be created to discourage change. Changing the way products are made or processes are operated is what pollution prevention is all about.

In order for pollution prevention to drive the development of new technologies, there must be predictability in the direction of the environmental regulations. There must be strong enforcement of the existing regulations and a demonstrable drive to pollution prevention as the highest level on the waste management hierarchy. None of these conditions are presently encouraging.

Environmental regulations have sunset provisions that provide for re-authorization within a fixed time period. Legislation is passed through extensive compromise thus making the outcome uncertain to all who follow the torturous path that these Congressional bills must follow. Once environmental legislation is passed, it is relegated to the EPA. The agency must then write new regulations or modify old ones. State governments, that can be more stringent than the new federal regulations, must scramble to be authorized to implement the new rules. All of this takes place in agencies with tremendous strains on personnel resources as exacerbated by Congressionally mandated time frames. Industries and environmental organizations then bring legal suits against the EPA in order to have the courts help provide an even different interpretation of these regulations.

Despite this climate of uncertainty, many public officials are strongly committed to pollution prevention. Speeches, Executive Orders, and other pronouncements have created a sense of anticipation that there is a priority for pollution prevention. However, these people must "walk the talk" to ensure that there is adequate follow through. One past EPA Regional Administrator made every department commit 15 percent of its budget to efforts to encourage pollution prevention. One state environmental agency took all the media-specific (e.g., air, water, hazardous waste, solid waste, etc.) programs and placed them into a Bureau of Waste Prevention. These moves were far more than symbolic gestures and demonstrate the support for pollution prevention at the highest levels.

There is still an approach to regulate industry one-pipe-at-a-time and to do this with end-of-the-pipe technologies. Some states have started to look at a multimedia approach giving the companies an opportunity to meet the requirements with prevention. However, federal regulations often simply prevent this. There is a growing trend to incorporate pollution prevention requirements separately in all media regulations. RCRA requires companies to have a waste minimization plan. In the stormwater program, a firm must have a stormwater pollution prevention plan. The TSCA program is developing a pollution prevention plan. Some sewage treatment plants require a pollution prevention plan for discharging wastewater under the pretreatment regulations. Oil storage often requires an SPCC Plan. Firms meeting the Clean Air Act requirements can file for an early emissions reduction plan to delay implementation of the regulations. Finally, many states are requiring plans. It is very difficult for firms to combine these disparate plans for fear that a regulator will be unable to determine compliance with the regulations.

Many companies have signed up with EPA's voluntary pollution prevention effort, the 33/50 Program. However, in most cases these firms have only exercised opportunistic changes in operating practices and materials substitution to attain their goals but lack a formal program to seek continuous improvement. There are often large variations in pollution prevention effectiveness in a multi-facility company. Most of this is attributable to the facility managers responsible for implementing these programs. Too often, the changes required are too great for them to be comfortable with.

A coatings manufacturing firm responds to its customers by developing new, "compliant" coatings. It is not unrealistic to assume that they will spend $2 million to $4 million to develop and test a new coating. It is likely that the cost of the new coating will be higher than that of the coating that it is designed to replace. In addition, the cost of the R&D has to be recouped in a reasonable period of time. Companies will pay more to remain in compliance with the regulations. However, there is not a great driving force to go to no-VOC coatings at a slightly higher cost if no one requires it. No one has gone to jail yet for not practicing pollution prevention. Thus, there is a greater risk for a coatings supplier seeking to fill the needs only of the firms who are seeking to stay out in front of the regulations.

As alluded to above, perhaps the greatest nonregulatory impediment is the resistance to change that exists in manufacturing. Many facility managers believe that processes are already being optimally operated. There is no perceived need to change technology. There is a lack of commitment to make this change even in today's competitive market. Often there is no written policy encouraging pollution prevention and no person is in charge of seeking prevention at the facility level.

When prevention programs are mandated by state and trade association initiatives, as well as EPA voluntary programs, there is a lack of personnel and resources assigned to the program. The environmental department's primary responsibility is to keep people out of jail for serious noncompliance. This is also a reactive situation like the historical perspective for manufacturing described above.

There is inadequate cost accounting for process losses. All environmental expenses are often placed in overhead and applied equally to all operations in the facility. However, some operations have greater environmental costs than others. Many facilities are turning to activity-based costing to remedy this problem.

In most firms there is no incentive system in place to reward pollution prevention efforts. Yet there are plenty of disincentives to punish noncompliant behavior.

Finally, there are intense production pressures that do not allow a lot of experimentation with the process. There is also a great concern regarding product quality. Other competitive pressures may also fuel a reluctance for companies to exchange information when technology changes lead to an increase in market share.

The best way to overcome these impediments and the reactive historical perspective is to foster the notion of continuous improvement in manufacturing firms. To some extent, many of these firms are already familiar with the concept of continuous improvement from their quality improvement initiatives. Instead of trying to focus on opportunistic targets in a pollution prevention program, a firm must identify opportunities for pollution prevention and have a program to move in the direction of new technologies that may bring breakthrough progress in the quest to minimize waste in the operation.

5. Opportunities for New Technologies

The focus on the difficulties involved in using new technologies for pollution prevention should not convince one that there is no need for process changes in a sustainable facility pollution prevention program. If the determination of alternatives for each pollution prevention opportunity is conducted properly, there will be a number of technological fixes for each case. However, if there is no generally applicable technology that can be developed to serve a particular functional need, it is likely that the facility will continue to modify and fine tune the process internally.

As more information becomes available in the Descriptive Approach format, a wide variety of technology needs will rise. However, some needs are universally needed at the present.

1. Surface Preparation and Coating Removal. This is a major consumptive use of chemicals. Halogenated solvents have been widely used for this application. Of course, many different surfaces and materials need to be removed from these surfaces. Many physical removal techniques may have wide applicability here. Industry has experimented with blasting with synthetic and natural materials (e.g., plastic beads or walnut shells); and use of Maxwellian light, hot and cold blasting media; and other energy-transforming means.
2. Surface Treatment. Companies are looking for less polluting technology for painting, plating, etching and cleaning. Substitute chemicals alone will not provide the complete answer. A means of using these materials effectively and efficiently in a closed system will be needed.
3. Improved Materials. This represents a very fruitful area for R&D and inventions. Corrosion-resistant materials would not need to be coated. Metal-free material substitutes are needed for inks, biocides, corrosion inhibitors, and paint. Substitutes for halogenated solvents are always in high demand along with alternatives to other toxic materials. Nonstick coatings would facilitate the cleaning of process equipment and other surfaces.
4. Improved Separations Technology. By using advanced membrane technologies, materials can be recovered and reused within the process. Advancements in this technology will even allow the recovery of dilute contaminants in high-volume waste streams.
5. Biotechnology-based Processes. Materials can be converted using microorganisms in lieu of chemicals. This often leads to lower energy use and reduces the use of toxic materials.
6. Process Monitoring Systems. Advances in the development of sensors will enable processes to be controlled in a manner so as to increase the efficiency of the operation.

These items pertain to a wide cross-section of industries. More examples will be provided in the industry-specific chapters in this book. It is important to understand where ideas that work in one industry can be applied to another. Increased demand for new technologies will go a long way to faster, further developments in this area.

6. Conclusions

A reluctance to change is firmly entrenched in manufacturing industrial facilities. Some of this is due to a historical perspective and some has been brought about by an uncertain regulatory and economic climate. However, this reactive stance is not conducive to remaining competitive in a global marketplace. Information on opportunities to apply technology to make manufacturing processes more efficient is needed for both the potential users and the suppliers of the technology to review. Whenever a process or operation is made more efficient, there will be fewer losses to the workplace and the environment as shown in the equation below:

$$\frac{\text{Materials Used} - \text{Materials Lost}}{\text{Materials Used}} \times 100\% = \text{Unit Process Efficiency}$$

This gain in efficiency is an important measure in the field of pollution prevention. Government must stop using its prescriptive checklists, questionnaires, and worksheets to get companies to conduct pollution prevention programs. Instead, the "Descriptive Method" which encourages firms to understand process functionality will be needed. There are few detailed process flow diagrams in the pollution prevention literature. If more information was available in this format, individual firms could utilize these maps to make refinements for the way they operate in their facilities. Technology vendors could utilize this basic information to identify industries where their inventions would be of use. In many cases, a number of different manufacturing industries have the same opportunity. Since the EPA often looks only at one industry at a time, these important cross-industry opportunities are often missed.

By holding focus groups with people from different industries that may have need for a given technology, vendors can get important ideas on how to position technologies for market entry. Vendors can also get some idea of the cultural issues that may be faced in certain industries and how impediments have been overcome in others.

By initiating partnerships between vendors and different industry groups, perhaps through a university, a particular pollution prevention technology can be piloted in a volunteer facility. This information can then be made available to others participating in the program. Because the firms are from different industries and not competing directly, this program can be quite effective. The competitors will find out about the change when the participants' market share begins to rise. Some forward-thinking trade associations may convince their members to share a particular pollution prevention technology that would benefit all members equally. Information on pollution control technology has been freely shared in the past within trade associations. There are ways to share pollution prevention technologies without antitrust violations. More efforts are required in this area.

Finally, there must be a push to continuously improve the maintenance of industry pollution prevention programs. The prescriptive approaches do not make this possible. By using the approaches described in this chapter, a firm can move steadily to the point where a technological approach will help increase the level of pollution prevention for a particular opportunity. By successfully easing into this technological alternative, the facility can overcome the resistance to change that has persisted for so long.

7. References

[1] R. B. Pojasek and L. J. Cali, Contrasting Approaches to Pollution Prevention Auditing, Pollution Prevention Review, 1(3), July 1991.

[2] R. B. Pojasek, Reviving the Feasibility Study for Use in Pollution Prevention Program, Pollution Prevention Review, 3(1), January 1993.

[3] R. B. Pojasek, Finding Safer Materials May Not Be as Easy As You Think, Pollution Prevention Review, 4(1), January 1994.

[4] National Research Council, Toward a New Era in U.S. Manufacturing, National Academy Press (Washington, D.C.) 1986.

[5] Business Roundtable, Facility-Level Pollution Prevention Benchmarking Study, Washington, D.C., November 1993.

About the Author: Dr. Robert B. Pojasek directs a national pollution prevention engineering practice with GEI Consultants, Inc., Winchester, Massachusetts. He helps industrial facilities plan and implement programs using quality improvement tools. He is also involved in reviewing pollution prevention technologies for venture capital groups. In addition to this work, Dr. Pojasek teaches a graduate-level course in pollution prevention at Tufts University, Medford, Massachusetts, and currently serves as Past President of the American Institute for Pollution Prevention.

Pollution Prevention in the Chemical Industry
Thomas W. Zosel
Manager, Pollution Prevention Programs
3M Corporation
St. Paul, Minnesota 55133
(612) 778-4805

A major difference between the chemical industry and most other industries is that the chemical industry *produces* chemicals while the others primarily *use* them. This makes it necessary to approach pollution prevention on a comprehensive scale, including both programmatic and technological elements in a program.

Increasing process yield and reducing waste production has long been a practice of the chemical industry. However, only within the last decade have these activities distinctly focused on pollution prevention. The chemical industry has always included source reduction, recycling, and reuse in its comprehensive, totally integrated approach to pollution prevention.

The pollution prevention technologies used by the chemical industry are as diverse as the industry members themselves. While a focus on yield improvement prevails, substantial efforts are now being directed at improving separation technologies, reducing the production of trace contaminants, and tailoring processes to recycle and reuse by-products and residuals.

Key Words

Adsorption; by-products; chemicals; inorganic; organic; polymerization; recycle; residuals; reuse; separation; toxicity; VOCs.

1. Chemical Users and Chemical Producers

Companies that are associated with chemicals can be divided into two classes: chemical *users* and chemical *producers*. Chemicals users are those that take the chemical, and without altering its chemical structure, use it during the manufacture of a product, as a constituent in that basic product, or as part of the manufacturing process.

An example of chemical use during the manufacturing process is the use of chlorofluorocarbons (CFCs) to clean circuit boards in the electronics industry. Here, CFCs never enter the actual composition of the product, but are nonetheless an integral part of the manufacturing process.

Chemicals are also used in the manufacture of pressure-sensitive adhesive tape. In this case, chemical organic solvents such as heptane, toluene, or xylene are used to dissolve the adhesive so that it can be applied in a thin layer to backing and manufactured into a pressure-sensitive tape construction.

Examples of chemical use as a constituent in a basic product include the manufacture of cosmetics and household cleaners. In both of these instances, various chemicals are blended in order to achieve product requirements that meet the customers' or users' expectations or needs.

In all of these situations substitutes for some of the chemicals used are possible. In the case of the electronics industry, CFC-free cleaning methods can be used to manufacture the product. These methods include water-based surfactants or other organic compounds which are less toxic or eliminate other adverse effects. The basic process can also be modified by using fluxless solder. This eliminates the cleaning step altogether. Details about the excellent work that has been done by the electronics industry is detailed in another section in this monograph.

In the area of cosmetics and household cleaners, the objective is to substitute materials that are less toxic but still satisfy the users. In the case of pressure-sensitive tape manufacture, the alternatives are to modify the process so that it recovers and reuses the solvent or completely eliminates the solvent by using a completely reformulated product. This can be accomplished through the use of a 100 percent solids hot melt adhesive, a solventless waterborne adhesive, or a two-part reactive system that produces the adhesive through an *insitu* polymerization process.

In all of these situations, a product which meets the functional needs of the customer can be produced using a completely different process or entirely different chemicals.

The other category of companies associated with chemicals are those that *produce* chemicals. These companies take raw materials and, through chemical processing and chemical reactions, produce a new chemical that itself is a product or that is used as a component to manufacture another product. In these cases, it is relatively difficult to develop an entirely new method of manufacturing a specific chemical. The chemical kinetics and stoichiometric relationships of chemical reactions are difficult to adjust or incompatible to the process of major change. While there are instances where major technological advances have drastically reduced the amount of waste generated by a specific chemical process, most pollution prevention achievements in the chemical industry have been accomplished through a series of incremental advancements. Consequently, the business process by which chemical users and producers approach the concept of pollution prevention can differ significantly.

This section focuses primarily on the approaches that chemical producers have used from both organizational and technological perspectives. In both instances, while significant accomplishments have been made, new management directions and new technological accomplishments are still needed to fully integrate pollution prevention throughout the industry as a whole.

The chemical industry is extremely diverse. It includes companies with billions of dollars in sales as well as those that have barely $100,000 in sales for their entire production. Consequently, some facilities that manufacture chemical products can have 5,000 or more employees at a site, while others may have fewer than 20. In these cases, the approaches used for pollution prevention can differ significantly.

The types of chemicals that can be produced are also diverse. Some facilities manufacture bulk chemicals. These facilities use a specific process to make tons of chemicals daily. An example is a facility that manufactures nylon for the carpet industry. In this case, an entire plant of over 1,000 employees may be solely engaged in manufacturing nylon yarn from basic chemicals. Another facility may produce no more than two or three pounds or even ounces of a specific specialty chemical on any specific day. It is relatively easy to imagine the significant differences in applying pollution prevention to a major nylon manufacturing facility versus a small specialty chemical manufacturing facility.

Another factor that differentiates the chemical industry is the manufacture of organic and inorganic chemicals. Organic manufacturing generally deals with chemicals that have a carbon backbone and may involve polymerization or use of solvents during the chemical reaction process. The wastes produced at these facilities almost always contain a high percentage of these organic materials. In addition, these facilities have wastewater treatment or pretreatment units that are highly specialized and must emphasize reducing air emissions of Volatile Organic Compounds (VOCs) from both process and fugitive sources. On the other hand, inorganic chemical manufacturing, which uses water solutions, generates acids and bases as by-products or wastes. These facilities present entirely different problems from both the pollution control and prevention perspectives. In general, inorganic chemical manufacturing produces large quantities of very specific solid wastes. Water pollution problems focus on acids and bases and air pollution focuses on particulates.

An important reality to remember in covering pollution prevention in the chemical industry is that chemicals are used throughout our society daily. Gasoline is an organic chemical that is manufactured through a depolymerization process called cracking of crude oil. Ethanol is a chemical that is manufactured through the fermentation of grains and sugar. The drugs that are used to combat disease are produced through complex chemical reactions from myriad constituents. Products that we use daily such as soaps, toothpaste, and deodorant are all chemicals that are manufactured through sometimes complex and other times simple chemical reactions.

It is also extremely important to understand that all chemicals are toxic. Every chemical that is produced and every chemical that we use in our daily lives can, depending on concentration and dose, be detrimental to the life and human welfare. Common table salt, if used to excess, can produce a toxic reaction. Many household cleaning products can cause toxic reactions if they are used inappropriately. Even laundry detergent can cause illness if ingested. Similarly, ethanol, if ingested in excessive quantities, can cause death. Yet having a beer during a baseball game is not viewed as a significant health risk. Consequently, when we refer to chemicals as toxic, we must also understand the concentration of those chemicals affects and the probability for adverse effects on human life and/or the environment. Therefore, it is important to understand the toxicity associated with a specific chemical and how that chemical is managed within the industrial environment. In that respect, it is extremely important to note that the chemical manufacturing industry has one of the best personnel safety records of any major manufacturing sector in the world. While the chemical industry handles many individual compounds that are regarded as toxic, they have established policies, procedures, and technologies that enable them to handle these materials safely and efficiently.

2. Pollution Prevention

2.1. History. Pollution prevention in the chemical industry is not a new concept. It has essentially been practiced since the first chemical processes were utilized in our industrial society. Initially, the industry focused on issues such as yield improvement rather than specifically preventing pollution from entering the environment. However, the consequence of both activities was the same: less material constituting waste streams entered the environment.

A good example of the length of time that pollution prevention has been practiced by a specific industry is given in Dr. Joel Hirschhorn's book, "Prosperity Without Pollution." [1] Dr. Hirschhorn references the sulfur dioxide emissions released from a sulfuric acid production facility in West Germany. The data presented clearly show that, over a long period starting in the late-1800s, this facility was implementing process changes that reduced the sulfur dioxide being emitted per pound of product produced. All of these manufacturing changes were essentially pollution prevention activities. However, only recently has the concept of pollution prevention been directly associated with these types of process changes, which were traditionally designated as yield or quality improvements.

The concept of chemical industry pollution prevention as we know it today began to emerge in the mid-1970s in response to the growing complexity and stringency of environmental requirements. Many proactive companies, notably some of those in the chemical industry, began to critically assess their approach to environmental issues. The concept of pollution prevention came through as clearly the most economical and environmentally effective means of addressing environmental challenges.

It is also extremely interesting to note that, from a historical perspective, the U.S. Environmental Protection Agency (U.S. EPA) first became interested in pollution prevention in the mid-1970s. As interest grew, U.S. EPA organized four regional conferences on pollution prevention in 1977. These conferences were cosponsored by the Department of Commerce and were held in Chicago, Boston, Dallas, and San Francisco. Each conference was organized by the separate U.S. EPA regions but in all cases, the U.S. EPA Administrator, either Russell Train or Doug Costell, participated as did the Secretary of Commerce, Mr. Elliott Richardson. While these conferences began to focus on pollution prevention, the concept was never really integrated into U.S. EPA's activities until many years later.

However, many companies and particularly those in the chemical industry, began to implement pollution prevention throughout their organizations through well organized, detailed programs. Perhaps the three best known programs are: 3M's Pollution Prevention Pays (3P Program); Chevron's Save Money and Reduce Toxics (SMART); and Dow Chemical's Waste Reduction Always Pays (WRAP). These three initiatives represent some of the first attempts by industry to integrate pollution prevention throughout an entire corporate structure. This approach has continued to grow to the point where U.S. EPA now espouses pollution prevention as one of its top priorities.

2.2. Definition of Pollution Prevention. While the formal concept of pollution prevention in the chemical industry has existed for almost 20 years, what exactly constitutes pollution prevention has only been an issue for the last few years since U.S. EPA and some state agencies have tried to address pollution prevention from a regulatory perspective. U.S. EPA's concept that pollution prevention is solely source reduction is not widely held in the industrial community. In fact, it is openly opposed by many of the successful practitioners from the chemical industry. Their concern is that by placing this limitation on pollution prevention, the significant environmental results of a fully integrated program will never be achieved.

The predominant view of the chemical industry is that recycling and reuse both on and off site should become an integral part of pollution prevention. Two overwhelming reasons account for this view. The first reason is that, for many companies, the goal of a pollution prevention program is to get as close to zero waste as technologically possible. In order to get to zero waste soley through source reduction, a chemical plant would have to make all of its processes 100-percent efficient. From a technological point of view, this is clearly impossible. Consequently, relying on source reduction alone will never allow companies to achieve the goal of zero waste. However, if recycling and reuse are included in the definition of pollution prevention, then every technical employee would be asked not only to make those processes as efficient as possible, but also to find a productive use for every by-product or residual stream. While that might be difficult, it is not technically impossible. Consequently, it can be viewed by both the employees and the company as a difficult, but achievable, challenge.

The second primary reason for including recycling and reuse in the definition of pollution prevention deals with the organizational aspects that drive entire organizations to a specific goal. In order to integrate a concept into a corporate culture, all employees must be able to incorporate pollution prevention into their daily job activities. If pollution prevention is limited solely to source reduction of toxic chemicals, then less than five percent of the employees of a corporation could participate in that pollution prevention initiative. Under those circumstances, a pollution prevention culture would never develop. However, if source reduction, recycling, and reuse of all wastes were incorporated into the program, then every employee could accomplish pollution prevention and help to integrate the concept throughout the entire corporate culture.

A recent benchmark study by The Business Roundtable [2] specifically addresses the definition of pollution prevention. The Business Roundtable conducted a study of six best-in-class facility pollution prevention initiatives. The objective of this study was to determine both the similarities and the uniqueness of each of these facilities' programs. In all cases, the facilities defined pollution prevention to be, at a minimum, source reduction, recycling, and reuse. In every case, the facilities had integrated their definition of pollution prevention throughout the entire organization so that every employee was responsible for the waste that they generated and for finding ways to actively reduce that waste.

2.3. Pollution Prevention Programs. In examining the state-of-the-art for pollution prevention in the chemical industry and the elements needed to move forward, two areas must be examined: pollution prevention *programs* and specific pollution prevention *technologies*. Pollution prevention programs are the organizational structures that are used to implement pollution prevention throughout an entire chemical manufacturing facility or an entire corporate organization. Both the overall pollution prevention program and the individual pollution prevention technologies are equally important in achieving environmental benefit. While the development of the technologies is an absolute requirement, without the organizational structure to support and implement those technologies, the environmental benefits that will result will never be achieved.

When chemical companies first began to address environmental issues, they focused on the installation of pollution control technologies, the classic end-of-the-pipe approach. In order to design and implement these controls, the companies hired a cadre of environmental professionals including environmental, civil, and chemical engineers with expertise in areas such as wastewater treatment, scrubber, and incinerator design.

However, as the focus changed from pollution *control* to pollution *prevention*, it became apparent that these environmental professionals did not have the knowledge or expertise to take maximum advantage of pollution prevention techniques. Those who could maximize pollution prevention were the scientists who invented the products, the engineers who designed the equipment, and the process experts who operated the manufacturing facilities. The primary challenge of any pollution prevention initiative was to enlist all of these individuals to focus a portion of their daily activities on reducing the environmental impacts of the operations under their direction. In short, the challenge was to obtain total corporate participation in pollution prevention.

The Business Roundtable [2] study to identify best-in-class facility pollution prevention programs highlights the elements that are necessary to fully integrate pollution prevention into a facility's operations. While this study points out the similarities between the various facilities' programs, it also emphasizes the fact that each program has many unique elements. Perhaps the most profound conclusion of this study was that each of the successful programs integrated pollution prevention into the existing corporate culture. They did not try to change the basic foundations of that corporate culture, but instead focused on how to adjust the existing corporate culture to fully accept the pollution prevention initiatives.

Successful programs shared five elements: top management support, involvement of all employees, recognition of accomplishments, transfer of information between facilities, and the measurement of results.

Top management support is necessity for a fully operational pollution prevention program. The top officials at both the corporate and facility levels must send a strong message to all employees that pollution prevention is an integral part of their jobs. This must begin at the level of the CEO of the corporation since that person sets the tone for all corporate activities. Chemical industry CEOs have been particularly active in this area. Within 3M, every CEO since 1975 has actively led the company pollution prevention program. Mr. Mahoney, CEO of Monsanto, was also instrumental in establishing the Monsanto Pledge which guides the company's response to all environmental issues. Likewise, Mr. Popoff, CEO of Dow Chemical, has been a leader in establishing sustainable development as a national goal as has Mr. Derr, CEO of Chevron. Mr. Woolard of DuPont has been the principal in incorporating pollution prevention into The Business Roundtable activities. In each case, these individuals have established pollution prevention as a top corporate priority. Each has promoted that concept within their own corporation as well as throughout the industrial community. All have established strong support through leadership by example.

While the facility's or corporation's environmental professionals may guide the pollution prevention program, the actual accomplishments occur through line operations. Involving all employees is critical to successful pollution prevention initiatives. Each employee knows their area of responsibility much better than the environmental professionals. While the environmental group is key in pointing out the problems, line operations are critical in developing and implementing the solutions. This cross functional effectiveness was a key finding of The Business Roundtable study.

Methods of recognizing and measuring pollution prevention accomplishments, were included in all of the best-in-class facilities' programs. Each designed programs that were consistent with their corporate culture and the goals that were established for the individual programs. Measuring accomplishments is extremely important since it represents a principal driving force for the employee's actions. Some facilities and corporate programs measure all wastes while other facilities and corporations focus on Toxic Release Inventory (TRI) emissions or on some other measurement methodology which best fits within their corporate culture and their pollution prevention programs.

It is clear that the programmatic aspects of pollution prevention have been a critical element in the environmental improvements made by the chemical industry. These programs establish the foundation that supports the development and implementation of the pollution prevention technologies. Without a well-established pollution prevention program, many good technical projects may never be implemented and a total commitment to the integration of pollution prevention technologies throughout the corporation and into every operating unit will be difficult, if not impossible, to achieve.

3. Pollution Prevention Technologies

The technologies that are used for pollution prevention in the chemical industry are as diverse as the chemical industry itself. Detailing all of these technologies and the needs for the future in a chapter is impossible. Consequently, this section is an overview not intended to paint a complete picture. However, it is possible to segregate current technologies into several specific categories: (1) the substitution of materials, (2) the modification of processes, (3) the reuse of material within the existing process, (4) recycling of material to a secondary process, and (5) the reuse of material in a different process. Each of these alternatives has various advantages and disadvantages as far as cost and implementation.

It may be worthwhile to briefly focus on the issue of pollution prevention costs and savings. Literature on the cost savings that can be accomplished through pollution prevention is extensive. For example, each year since 1975, 3M has calculated the savings that has been accomplished through its Pollution Prevention Pays Program. In the 18-year program, the savings have exceeded $700 million. However, the important thing to note is that projects included in that program are those that actually save money. A significant number of other projects have been proposed, but not implemented because, while they were pollution prevention projects, they did not offer savings.

This attribute of successful pollution prevention programs is supported by the information that was contained in The Business Roundtable study [2]. All of these successful programs implemented projects that both prevented pollution and saved money. Each of the facilities had projects that focused on pollution prevention that were not implemented because they were not cost effective. Consequently, the common belief that all pollution prevention projects will save money is not accurate. However, it can be said that all successful pollution prevention initiatives develop and implement projects that are cost effective. Consequently, cost savings associated with a project are integral to that project's successful implementation.

In looking at the state-of-the-art existing technologies, the American Institute of Chemical Engineer's Center for Waste Reduction Technologies (CWRT), along with the Department of Energy, conducted a comprehensive study which culminated in a workshop to determine what pollution prevention technology areas are needed for the chemical industry. Many of the suggestions for future research or areas that need specific development effort from this CWRT initiative are discussed in the following sections.

3.1. Material Substitution. While chemical users have relied heavily on materials substitution to achieve pollution prevention, this alternative is significantly more difficult for chemical producers to achieve. To produce a specific chemical, specific reactants are necessary. In order to produce that same chemical with different reactants, an extensive research effort into the basic pathway of the chemical's formation is necessary. Even then, it is probable that even significant research may fail to identify an alternative pathway.

An example is nylon production. Nylon is a polymeric material produced through the chemical reaction of adipic acid and hexamethyldiamine. Individually, these compounds are more toxic than the end product, since nylon is quite biologically inert. If the objective was to reduce the use of adipic acid or hexamethyldiamine, then a completely different chemical reaction pathway for the production of nylon would need to be developed. In many cases that involve similar operations, it is already well known that alternative pathways either do not exist, produce products which are inferior, or are otherwise environmentally disadvantageous.

Notwithstanding the difficulties inherent in material substitution, the chemical manufacturing industry has continued to pursue this course. Efforts have been directed more toward the production of substitutes for existing chemicals than the constituents that are used to manufacture the chemicals in the first place.

An example of this substitution would be in the area of CFCs. Because of their impact on stratospheric ozone, CFCs are being phased out through the requirements of both the Montreal Protocol and the Clean Air Act Amendments of 1990. This phase-out caused the chemical industry to develop chemicals without stratospheric ozone depleting potential to replace CFCs. While this effort is far from complete, chemical alternatives have surfaced as CFC replacements.

When examining chemical replacements, it is important to understand that many of the attributes which make the chemicals ideal for their intended uses may also contribute to the problems that they create. CFCs, for instance, are nonflammable and somewhat chemically inert although they do have some excellent solubility characteristics. The chemical bonds between the carbon and the fluorine and chlorine make them very stable. This allows CFCs to, over time, migrate to the stratosphere where, under heavy ultraviolet radiation, the bonds between the carbon and chlorine are broken. This creates the chlorine-free radical which is the responsible element in the reaction with stratospheric ozone.

Consequently, a chemical replacement for the CFCs needs to meet the same chemical activity characteristics yet eliminate the creation of the chlorine-free radical in the upper atmosphere. Perfluorinated carbons (PFCs) are being investigated as a replacement for CFCs. These materials contain no chlorine. All of the halogen bonding occurs between the central carbon atoms and fluorine. The carbon-fluorine bond is much stronger than the carbon-chlorine bond, thus making these compounds even more stable than CFCs and substantially reducing the ability for UV radiation to attack the bond to create free radicals. Currently, one of the drawbacks to the use of PFCs as a more widespread CFC replacement is that it costs more to produce.

While other examples of material substitutions exist in the chemical industry, there is no common thread in the technological direction of these changes or in the research that is needed to facilitate them. Because of each chemical's unique characteristics and the specific chemical reactions that are needed to create it, each case of chemical substitution in the chemical industry must be dealt with individually. Therefore, it is impossible to specifically delineate the types of future reaction or kinetic research that will yield the greatest opportunities to develop new chemicals to replace existing chemicals with negative environmental attributes.

3.2. Process Modifications. While it may be difficult to delineate the technological developments needed for development of substitute materials, efforts to develop technologies or process innovations that will facilitate pollution prevention are proceeding. Most of these process changes aim either to improve the yield of the chemical reaction (i.e., produce more product from the given set of raw materials) or to eliminate or reduce unwanted reaction products (i.e., contaminants that result in waste or that need to be removed, which again generates waste).

In regard to basic reactor design, an analysis conducted by CWRT revealed that the development of uniform conditions within a reactor is an area for further research. Because of variability in heating and/or mixing, variations in temperatures or concentrations may occur as a reaction proceeds within the reactor. By making those conditions more uniform throughout the entire reactor, chemical reactions would likewise be more uniform. Hopefully, this would increase the yield of the chemical to be produced while also reducing the amount of by-products to be removed before the product was available for its intended use.

Several research areas can address this specific need. The most fruitful are the materials of construction and the configuration of the reaction vessel. With the advent of higher-strength alloys that resist corrosion, the ability to develop creative designs for reaction vessels has significantly increased. There has also been a trend away from the batch reactor toward the use of continuous flow reactors in some situations. These systems allow greater control of the temperature and mixing within the flow of materials, but necessarily are limited to high-volume production levels. Establishing a continuous flow reactor design for low-volume production would not be practical. It has been asserted that side reactions can occur on the actual surface of the reactor vessel. This could occur either because the surface material was acting as a catalyst for a reaction that was competing with the intended reaction or because the competing reaction was occurring because of a surface-initiated phenomenon.

It is important to note that this competing reaction need not produce a significant quantity of material or cause production problems to be environmentally undesirable. Trace contaminants produced in this manner may be easily separated from the intended output. However, the process necessary to accomplish this separation can generate significant wastes and environmental contaminants. Consequently, eliminating the trace contaminants avoids the process needed to remove them. As emphasis on reducing the environmental impact of the chemical industry has grown, projects such as this have gained far greater importance.

A second area delineated by CWRT for future research is surface kinetics. This includes examining whether reactions that occur on the surface of the chemical reactor play a primary role in adding to the trace contaminants within that chemical process. The objective of the research is to develop reactor design and kinetics that improve yield and eliminate the production of trace contaminants that must be removed in subsequent processing that produces by-products or wastes. Primary research needed in this area would promote a better understanding of how surface phenomena can impact a chemical reaction or change the kinetics in even a very minor way so as to generate an unintended by-product or to increase the production of a by-product that would be included in the intended reaction.

There is also a need for research to determine if the materials of design can act as catalysts to facilitate side reactions or to promote competing reactions that would generate trace contaminants or affect product quality. At 3M, a scale-up of a new water-based adhesive from the pilot to the production level was a significant problem until it was discovered that, while the reaction vessel was glass lined, one of the newly installed raw material piping systems included a section of iron pipe. The extremely small quantity of iron that was introduced into the reaction vessel as the raw material flowed through this iron pipe was sufficient to create quality problems in the adhesive. This produced a significant quantity of waste because it was necessary to filter the adhesive. Once the source of the problem (i.e., the trace contaminant of iron) was identified and eliminated, the amount of waste generated by the filtering system was significantly reduced.

Staying with the reactor design, the CWRT analysis also showed a need for rapid quenching of the reaction products. Once a reaction has been completed, it is necessary to stop that reaction and remove the materials from the reactor. If this is not done quickly, portions of the reaction which degrade the original compound may proceed producing trace contaminants which must be removed later in the processing of that chemical. Consequently, there is a significant need within the industry to develop methods to shut off reactions. One of those methods may be rapid quenching of the material exiting the reactor or some other innovative way of stopping the reaction kinetics from proceeding.

Another broad area of need in process research is improved separation techniques. A significant amount of waste from the chemical industry is created because trace contaminants or by-products must be removed from a specific chemical stream. If that cannot be done through reactor design or reaction kinetics, improved ways of separating those trace materials to prevent production of significant amounts of contaminants are needed. This extremely broad area of research offers many opportunities for new and unique work in the area of separations.

Separation processes are used to increase the purity of the intended product. This can be as simple as decanting immiscible liquids or as difficult as trying to break an azeotropic mixture of solvents or removing a trace contaminant that is present in the parts per million range. The types of processes that are currently used by the chemical industry to accomplish this vary widely and include filtering, distillation, adsorption, absorption, steam stripping, etc.

The important point for this discussion is that all separation processes generate either a waste or a by-product. The term "waste" refers to a residual that must be treated before disposal or that goes directly to disposal. A by-product, on the other hand, is a material that is produced at the chemical plant which is not the intended product, but is to be returned to commerce.

An example by-product is the methanol produced during the polymerization of dimethyl terephthlate and triethylene glycol to produce a polyester resin. This methanol must then be purified through a separation process to remove the contaminants which are the waste and produce the methanol which is the by-product.

Consequently, if separation processes can be made more efficient and effective, the amount of waste material will be reduced and the amount of the intended product and by-product will be increased. Because of the extremely wide range of separation processes which are used in the chemical industry, it is difficult to outline all of the areas that are ripe for new technological advances. The standard unit operations such as distillation or stripping are extremely well studied. While advancements in these operations are possible, they are not the most productive areas for future focused research.

On the other hand, some excellent work is being done in areas including specialized filtration, adsorption, absorption, and membrane technologies. The work in the area of semi-permeable membranes is particularly promising when viewed from the environmental perspective. The basic concept is to create a membrane that will allow some materials to pass through while being impermeable to other compounds. In most cases, a pressure gradient or a concentration gradient is the driving force that moves the material through the semi-permeable membrane. Based on this technology, several reverse osmosis applications have been highly successful in the wastewater area.

However, even more innovative work is being done in the air area. One of the most pervasive environmental challenges to the chemical industry is the separation of low-concentration VOCs from air streams. Many chemical processes, particularly in the manufacture of organic chemicals and petrochemicals, result in low concentrations of VOCs in an air stream. In these situations, the semi-permeable membrane can work in two different ways. One allows the VOCs to pass through while remaining impermeable to the air which acts as a carrier for the VOC. This system is necessary for high air flows and is the area where most research is currently being conducted. The opposite situation, allowing air to pass through while remaining impermeable to VOCs, can potentially be used for low flow streams of high VOC content. To date, there has not been a significant amount of work directed in this area.

There is also the opportunity to design a semi-permeable with the potential to selectively pass specific VOCs. In the area of separation technology, this would be a significant advancement. It would allow either the removal of a trace contaminant from an air stream that, with the contaminant present, could not be used as intended, or it could remove a low-concentration VOC product from an air stream in a cost-effective manner.

Another area of separation technology is the development of media that can selectively remove an individual compound or a group of compounds from either a gaseous or liquid stream. This can be accomplished either through adsorption or absorption. The use of activated carbon to separate general organic compounds from air or water streams has been well recognized and has been an integral part of chemical industry operations for years.

Recent research in this area has focused on allowing carbon adsorption separations to occur in a much more environmentally responsible manner. The historic use of carbon adsorption to separate VOCs from an air stream involved the steam stripping of the activated carbon to remove the VOC for subsequent use. If that VOC was water miscible, a subsequent distillation process was required to remove the VOC from the condensed steam/VOC mixture. If the VOC was not water miscible, it would easily be decanted after the steam was condensed. However, the water phase would need to be treated wastewater because of trace contaminants from the VOC.

Newer carbon adsorption installations are now focusing on a hot inert gas stream as the desorption media with subsequent condensation in a chilling system or through a compression-expansion cycle. The use of inert gas eliminates the explosion problem that could occur if activated carbon containing a flammable VOC was desorbed into an air stream. The inert gas is generally either nitrogen or a gas which is produced by burning a low molecular weight hydrocarbon such as propane under stoichiometric conditions so that less than five percent oxygen is present in the combustion product. This gas then must be dried to reduce the water vapor that is produced.

After the carbon has been desorped, the resultant gas stream is put through a chilled condenser which is below the dew point of the VOC in the inert gas. This causes the VOC to condense and, consequently, it is removed as a liquid. An innovative twist on this method is to use a compression-expansion cycle for cooling. The first step is to compress the VOC-inert gas mixture. This generates heat which is then removed from the compressed gas. This compressed gas is then allowed to quickly expand which causes cooling. If the cooling is below the dew point of the VOC in the inert gas, the VOC will condense and can be separated as a liquid.

Another recent advance in the separation of VOCs from water is the use of polypropylene micro fibers. These fibers are hydrophobic and oleophilic. In simple words, they hate water and love oil. Consequently, these fibers can be used to separate oil or any oil-like hydrocarbon from a mixture with water. The organic material can then be removed from the fiber by a variety of physical methods. In many cases, physical pressure, such as running the organic saturated fiber through a set of nip rollers, is sufficient.

3.3. Recycle and Reuse. The recycling and reuse of materials are extremely important in the chemical industry's approach to pollution prevention. As discussed earlier in this section, it is impossible to make every chemical process 100 percent efficient, (i.e., convert all raw materials to the intended product). Consequently, a chemical manufacturing facility will always have residual or by-product streams that must be addressed through pollution prevention.

In the past, most companies focused solely on increasing the production of the intended product. Residuals or by-products produced were disposed of as secondary materials. As the focus on pollution prevention has become paramount to many chemical operations, people have realized that tailoring a process to produce by-products that are recyclable or reusable commodities rather than wastes can be environmentally effective and cost efficient.

An example is the conversion of processes that used carbon adsorption systems to separate VOCs from air streams to direct condensation where applicable. While carbon adsorption is still very effective, under certain circumstances, the surface phenomenon associated with the adsorption process can facilitate chemical reactions that produce trace amounts of contaminants. If this occurs, the recovered organic material must be subjected to a secondary purification process such as distillation, before it can be effectively reused.

However, if rather than recover the by-product VOC through a carbon adsorption system, the company converts to a condensation system, the VOC can be recovered without the formation of the trace contaminants. This would allow VOC to be reused directly rather than go through the secondary purification processes. This change can substantially reduce both the amount of waste that is produced and the amount of energy that is used for recovery and purification.

Additional research must focus on abilities to recycle and recover materials in specific waste streams. State-of-the-art methods, particularly through condensation or carbon absorption, systems, allow recovery of relatively high concentrations of organics in air streams. However, these methods do not allow for the economic recovery of VOCs from very dilute streams and particularly very dilute streams with high moisture content. Consequently, it has been suggested that research focus on the ability to recover VOCs from dilute streams that contain moisture. Many of the streams emanate from chemical facilities, and particularly from those which produce organic chemicals. Consequently, this area of research could be very fruitful for future pollution prevention projects.

In discussing the recycling and reuse of materials, separation technologies are extremely important. Because the removal of trace contaminants and the separation of the by-product so that it can be reused or recycled is critically important.

In determining which technology should be used for a by-product that is to be reused or recycled, it is important to understand not only the capabilities and limitations of the process, but also how the recycled or reused material will be used. As an example, there are two gold mines in Nevada. One mine is installing a roasting process for a sulfite ore that will produce significant quantities of sulfur which can be converted into sulfuric acid. A nearby mine is utilizing an extraction process which requires a significant amount of sulfuric acid in order to effectively extract the gold. This creates a symbiotic relationship between the two facilities in that what would have been a waste from one is converted to a raw material for the other.

However, in developing the technology that will be used at the first mine, the strength and purity of the sulfuric acid that will be needed at the second mine must be considered. Since the sulfuric acid is to be used in an ore extraction process rather than in a chemical process, a reagent grade sulfuric acid is not needed. Consequently, the process design at the first mine does not need to include separation processes that are designed for a high-purity product. The same would hold true for the concentration. The mine using the sulfuric acid would like the produce to be as concentrated as possible without adding significantly to the production and transportation costs. The major point illustrated by this example is that, when designing processes or technologies for the reuse and recycling of materials, the way that material is to be used must be considered.

Technologies that allow for total water reuse within a process or within an entire chemical facility are also needed. A significant amount of water is used by chemical industry facilities. This water becomes effluents that carry contaminants into the environment. If ways were found to allow those facilities to become totally closed systems, (i.e., where all the water would be continually reused within that facility), then a significant amount of pollutants would be prevented from entering the environment.

In order to accomplish total water reuse, several areas needed attention. One area is the ability for trace contaminant removal. If water is to be reused, the trace amounts of chemicals must be removed before re-injection into the process. Removing those trace contaminants has become an environmental problem. It would be a similar problem if that would be used within that facility. Consequently, ways of removing trace contaminants through adsorption, absorption, or reaction are fruitful areas for research.

The CWRT also highlighted the need for focused component reactions in which, rather than remove the contaminant through a removal technology, subject the water with trace contaminants to a chemical reaction. That chemical reaction would eliminate those trace contaminants and allow them to be easily removed from the wastewater stream. This is an area which has not been extensively researched.

Another area of research that was suggested in this analysis was to develop methods of organic neutralization so that organic acids and bases could be neutralized without using extensive amounts of water and inorganic acids as is currently done. The use of water and inorganic acids produces a significant amount of waste for various processes within the chemical industry. Consequently, new technologies that allow direct organic neutralization would significantly reduce waste from various processes within the chemical industry.

This section represents only the tip of the iceberg concerning technologies now in use for pollution prevention in the chemical industry and those technologies with significant potential for future application. As delineated at this section's beginning, the chemical industry is extremely diverse, consisting of both very large and very small companies and facilities. There is both organic and inorganic manufacturing. But, while there is a constructive diversity, there is also a strong environmental ethic which places pollution prevention (i.e., source reduction, recycling, and reuse) as a top industry priority.

4. References

[1] Joel S. Hirschhorn and Kirsten U. Oldenburg, Prosperity Without Pollution, Van Nostrand Reinhold, New York, 1991.

[2] Facility Level Pollution Prevention Benchmarking Study, The Business Roundtable, November 1993.

About the Author

Thomas W. Zosel is the manager of Pollution Prevention Programs at 3M Corporation.

Pollution Prevention in the Electronic and Office Equipment Industry

Patricia A. Calkins
Corporate Manager of Resource Conservation
Jack C. Azar
Manager of Environmental Design and Resource Conservation
Xerox Corporation
Webster, New York 14580
(716) 422-9506

Office equipment and electronic products present some unique and important end-of-life issues related to both hazardous and solid wastes. Increasing attention to these issues has spurred pollution prevention efforts aimed at two primary objectives: (1) reducing the amount of hazardous materials, and (2) extending the useful life of such products.

The future life-cycle vision for the electronics and office equipment industry is to have a closed-loop system including reuse, remanufacture of equipment, and materials recycling. Each step in this system presents technical and business challenges and opportunities including:

      An equipment collection infrastructure
      Disassembly and parts/materials sortation
      Parts refurbish/remanufacture
      Recycling technologies
      Materials science issues
      Development of new design tools
      Development of new design technologies

In this chapter, each of the above categories is individually discussed. In addition to identifying technology needs, examples of appropriate solutions are indicated.

Key Words

Disassembly; electronic; hazardous materials; life-cycle; office equipment; plastics; pollution prevention; recycling; refurbish; remanufacture; separation; sorting.

1. Introduction

1.1. Regulatory Thrust Shifting to Products

1.1.1. Pollution Prevention Efforts Address Product End-of-Life.
Office equipment and electronic products contain critical materials and components that, at end-of-life, are treated as hazardous waste (e.g., circuit boards containing lead; rechargeable batteries containing cadmium, nickel, and mercury; cathode ray tubes containing leaded glass). This, combined with an increasing solid waste problem, has stimulated concern for product end-of-life management. As a result, worldwide regulatory activity has escalated rapidly.

This drive toward manufacturer product stewardship (or producer responsibility) is happening in Germany, The Netherlands, Denmark, Austria, Sweden, and Switzerland. Either through legislative or voluntary means, these countries, as well as the United States, will eventually push product end-of-life responsibility back up the chain ultimately to the equipment supplier. In response, the electronic and office equipment industry is directing pollution prevention efforts toward product end-of-life management.

Pollution prevention, as applied to end-of-life office and electronic equipment, involves two primary objectives:

> Reducing the amount of hazardous materials used by such products by developing alternative materials that are less toxic to humans and safe for our eco-system (i.e., clean technologies).
> Extending the useful life of such products and components to minimize their environmental impact. Product life extension can occur either for the same application or for other related uses.

1.1.2. Products Currently Managed as Waste at End-of-Life.
This new focus on products is based on the realization that manufacturing is not the only stage of a product's life that affects the environment. Many products end up in the waste stream after serving their useful life. While quantitative information in the United States does not exist, in France, electrical and electronic equipment products are reaching end-of-life at a current rate of nearly 1.3 million tons per year and steadily increasing [1]. Considering that the U.S. population is 4-5 times larger than that of France, our annual electrical and electronic waste volume could exceed 6 million tons. Reversing this trend requires developing an entirely new infrastructure with a reverse distribution system to get back machines and technologies that will enable used equipment contents to be recovered as valuable assets.

In the fall of 1993, a group of professionals representing more than a dozen electronics and office equipment companies and recyclers assembled to collectively discuss their views on managing products at end-of-life [2]. They determined that an inadequate collection and separation system was making equipment and materials retrieval very difficult. Additionally, once collected and returned, they faced the dilemma of what to do with the recovered equipment. In most cases, landfill is the only disposition option available.

1.1.3. Companies Agree an Infrastructure Must Be Developed.
An overwhelming majority agreed that to effectively reuse, remanufacture, and recycle office equipment, an equipment reverse distribution system is needed and a recycling infrastructure must be developed. Improving this process requires a dual approach that incorporates an end-of-life value into the design of future products *and* develops technologies that help maximize the value of equipment currently reaching end-of-life. The more specific technical and logistical elements for both of these approaches are covered throughout this chapter.

2. Vision

2.1. Working Toward a Closed Loop. The future vision of the electronics and office equipment industry is to have a completely closed-loop system. As illustrated in Fig. 1, the process for working toward this vision includes reuse, remanufacture, and material recycling. As noted previously, the first step in the entire process is to get back equipment. Once retrieved, the equipment can then be separated into that which can be directly remanufactured and marketed and that which is destined for disassembly. The contents of disassembled equipment fall into two general categories; refurbishable parts and recyclable materials. Ideally, refurbished parts go directly back into new equipment and recyclable materials become new parts. Where this is not possible, other applications outside the industry are sought.

While some office and electronic equipment companiés currently remanufacture their products on a limited scale, there remains great opportunity for enhancement. As noted in Fig. 2, remanufacturing is a fairly complex process consisting of many unit operations. While traditionally performed in-house, to effectively optimize remanufacture and maximize materials recycling, external partnerships will need to be developed (See Fig. 3).

Each step along this process presents opportunities for small/medium sized businesses. The following needs in the electronics and office equipment industry are discussed in this chapter:

        Equipment collection infrastructure
        Equipment disassembly and parts/materials sortation
        Parts refurbish/remanufacture
        Material recycling technologies plastics
        Material science
        Design tools
        Design technologies
        Market pull
        Other issues

3. Approach

3.1. Getting Products Back is the First Step

3.1.1. Issues. Issues related to recovering used equipment are essentially influenced by how the equipment is marketed and its perceived (or real) value. For the most part, this can be related to equipment size and purchase cost. Office equipment suppliers have fairly good control over recovering larger, more expensive, machines. On the other hand, smaller machines have been traditionally more difficult to recover [1]. While equipment recovery issues are most relevant to smaller machines, parts and material recovery apply to all machines.

Larger machines are generally direct marketed by the manufacturer to commercial and corporate customers. Often, old equipment is replaced with new by the same manufacturer. Additionally, these machines tend to contain more valuable re-usable parts and materials and can be expensive to decommission. In these cases, the office equipment supplier has an opportunity (and frequently the contractual responsibility) to take back ownership of the used equipment (as illustrated in Fig. 4).

In contrast, smaller machines such as faxes, small copiers, personal computers, and low-end printers are typically distributed via retail channels to consumer markets. As shown in Fig. 5, at end-of-life, users perceive no scrap value and, with no convenient mechanism for getting rid of old machines, the equipment ends up in the municipal waste stream. Additionally, the sheer size of the United States presents a significant logistical problem for collection.

3.1.2. Opportunities. The primary opportunity here is probably more logistical than technical. However, it is important to understand that it is critical to address this step before subsequent technologically based solutions can be applied (e.g., disassembly technology is useless without machines to disassemble). The collection infrastructure can either be developed from scratch or possibly by capitalizing on existing distribution and collection systems associated with other industries.

Ideally, this type of operation would be able to collect any type of electronic and office equipment regardless of the manufacturer. From here, the process can follow two potentially different routes or a combination of the two. Collected equipment can be sorted and returned to the original manufacturers; or it can be disassembled to varying degrees as described in subsequent sections of this chapter. Disassembled components can in turn be redistributed to new markets or be reprocessed and incorporated back into the same equipment.

> Capitalizing on existing distribution systems, the business opportunity could involve agreements with both retailers and equipment manufacturers. As a service to their customers, retailers (or their contractors) would take back obsolete and non-functional equipment. A third-party company would collect returned equipment from retailers. This equipment would then be sorted according to manufacturer. When a predetermined number of units was collected, they would be returned to the original manufacturer or sent to an industry dismantler.
>
> Municipal waste collection companies are an example of an existing collection system that may present an opportunity for recovering end-of-life equipment. These companies could be contracted by an equipment recovery company to include consumer electronics collection in their process. Alternatively, municipal waste companies would expand their own service to include such collection. For example, Laidlaw Waste Services Ltd. in Ontario is developing a process to recover and process electronic equipment.

Once collected, small/medium size business may want to add processing capabilities to address the issues identified later in the chapter.

3.2. Disassembly and Sortation Designed to Maximize Used Part and Material Value

3.2.1. Issues. After equipment has been retrieved, subsequent steps in the process must be designed to maximize content value of machines that cannot be remanufactured. Recoverable equipment contents fall into two general categories: reusable parts and scrap materials. A wide range of parts are potentially reusable including equipment frames, covers, motors, circuit boards, computer chips, etc. Metals and plastics comprise a significant proportion of scrap materials having potential recycling value.

Even companies with well-established remanufacturing processes find that many parts end up as waste scrap. Metal parts are generally dispositioned into scrap metal recovery streams; other materials are discarded as waste. As discussed later, and depicted in Figure 6, with proper technologies and a logistical infrastructure, many parts can be refurbished for reuse and other materials can be recycled. Effective recovery depends on cost-effective disassembly and sorting without damage to valuable parts.

For the most part, disassembly and sorting is currently accomplished using labor-intensive manual operations. Moreover, due to traditional attachment technologies (e.g. welding, soldering, adhesives) many parts are virtually impossible to separate without damaging functional integrity. Consequently, current separation and segregation processes are difficult and costly.

3.2.2. Opportunities. The following areas represent opportunities for pollution prevention activities related to disassembly and sortation.

Processes to sort and separate parts and materials cost effectively with proper removal and disposal of hazardous wastes. Current disposition processes rely on limited information supplied by OEMs as to the materials contained in recovered products many of which are obsolete. As a result, costs for identifying, removing, and properly dispositioning materials can be high. Processes are often very inefficient which discourages recyclers and OEMs from investing in technology and applying resources to this problem. Developing hazardous materials data bases for materials commonly used in office equipment including application and proper dispositioning procedures, would certainly assist small businesses and entrepreneurs in such activity.

Bar coding systems that help sort and track the history of individual parts will enable effective life cycle utilization and appropriate dispositioning of parts and materials [3].

Rapid part/material identification (automation where appropriate). Technologies for automated non-destructive identification of parts and engineering materials have not as yet been commercialized. While high-speed identification of plastic packaging has been achieved and is commercialized, rapid recognition of engineering plastics, which are utilized most in business equipment, is not commercial. Only with such recognition systems will high-speed, low-cost parts sortation be achievable.

The primary technologies currently known for this application include near infrared (NIR), mid-range infrared (diode lasers), and microwave interaction plastics sensor (MIPS). These technologies are all currently in development and each is associated with unique problems [4]. With continued development, these technologies should be marketable within the next two years.

Design-for-disassembly technologies to enable rapid and cost-effective parts recovery for reuse/remanufacture. These design technologies are currently being pursued in both academia and private industry. Carnegie Mellon University is building knowledge in this area within their mechanical engineering department. IBM and Xerox are developing expertise within their respective design communities [5,6]. Automated disassembly techniques may be effective for very high-volume products such as personal computers and consumer electronics while manual disassembly technologies for lower-volume products need to be easy and efficient.

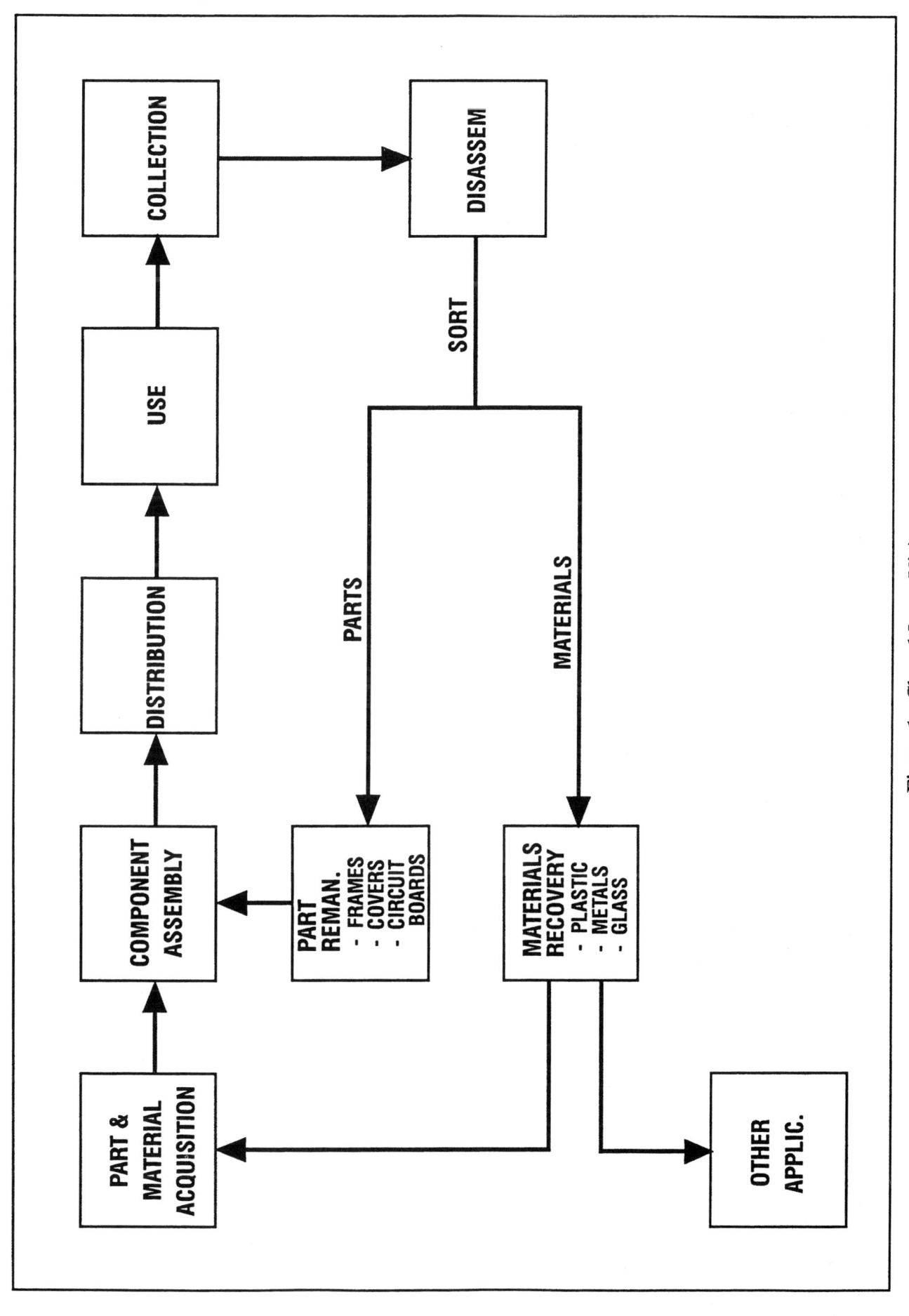

Figure 1. Closed-Loop Vision

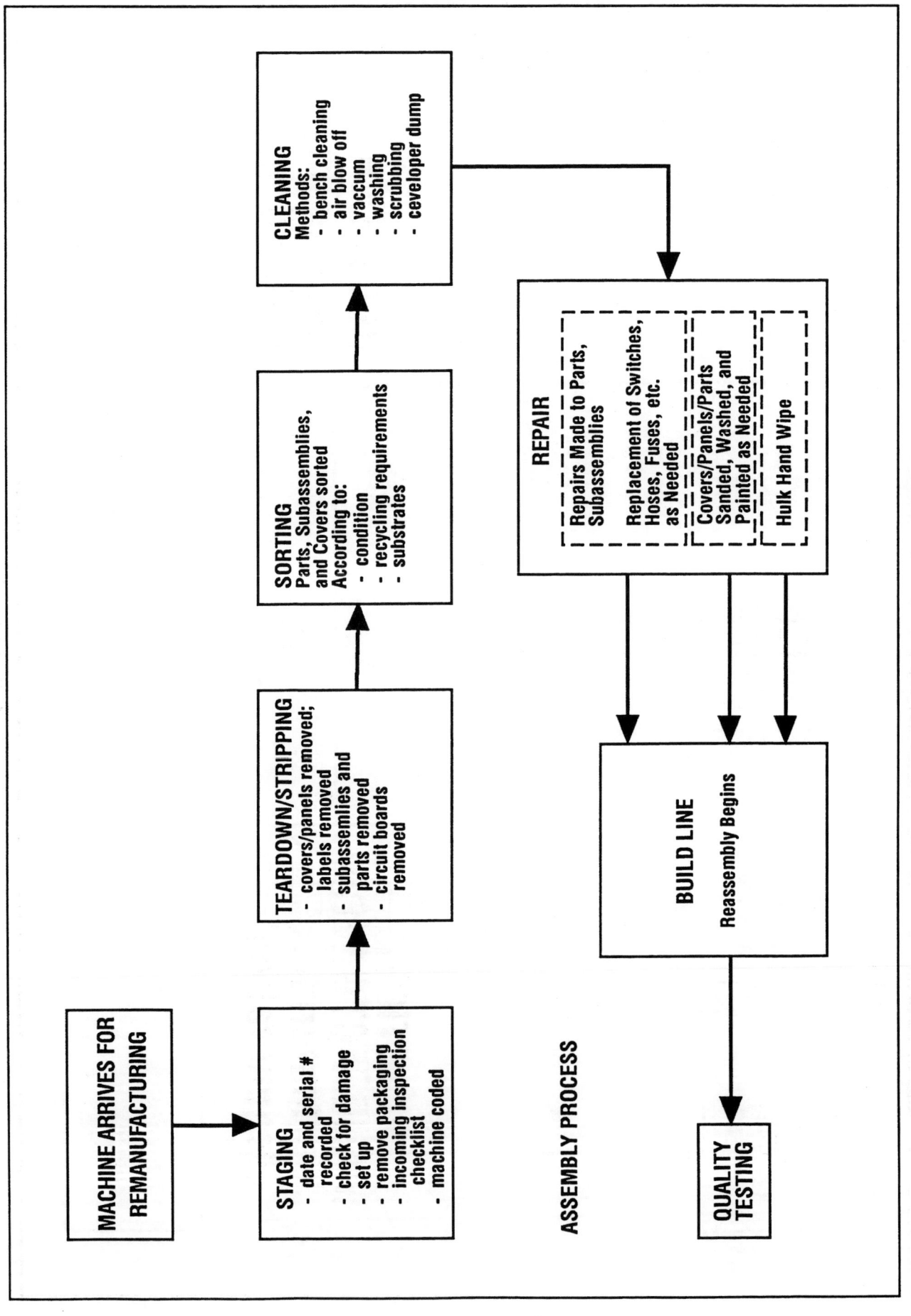

Figure 2. Typical Remanufacturing Process

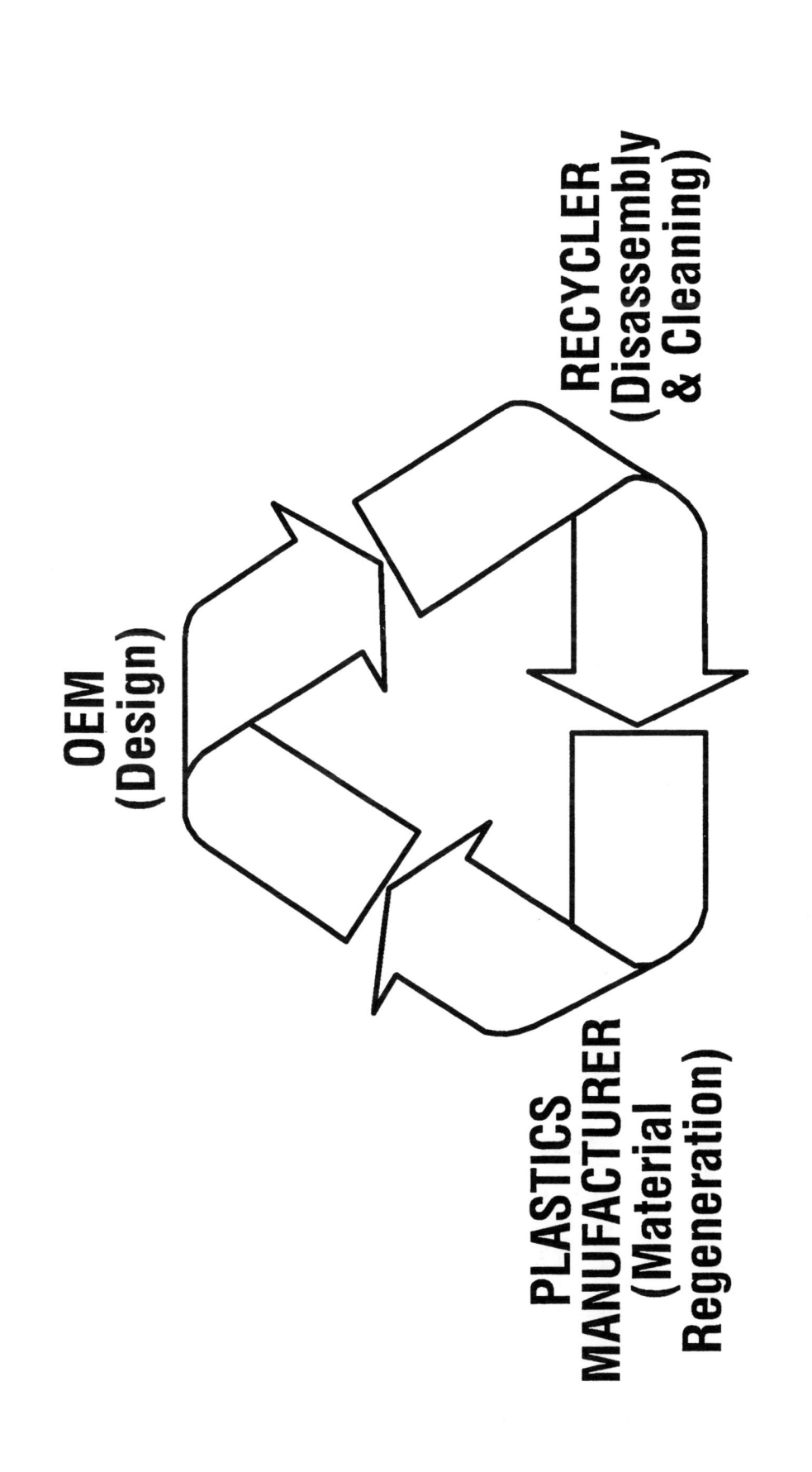

Figure 3. Closed-Loop Recycling

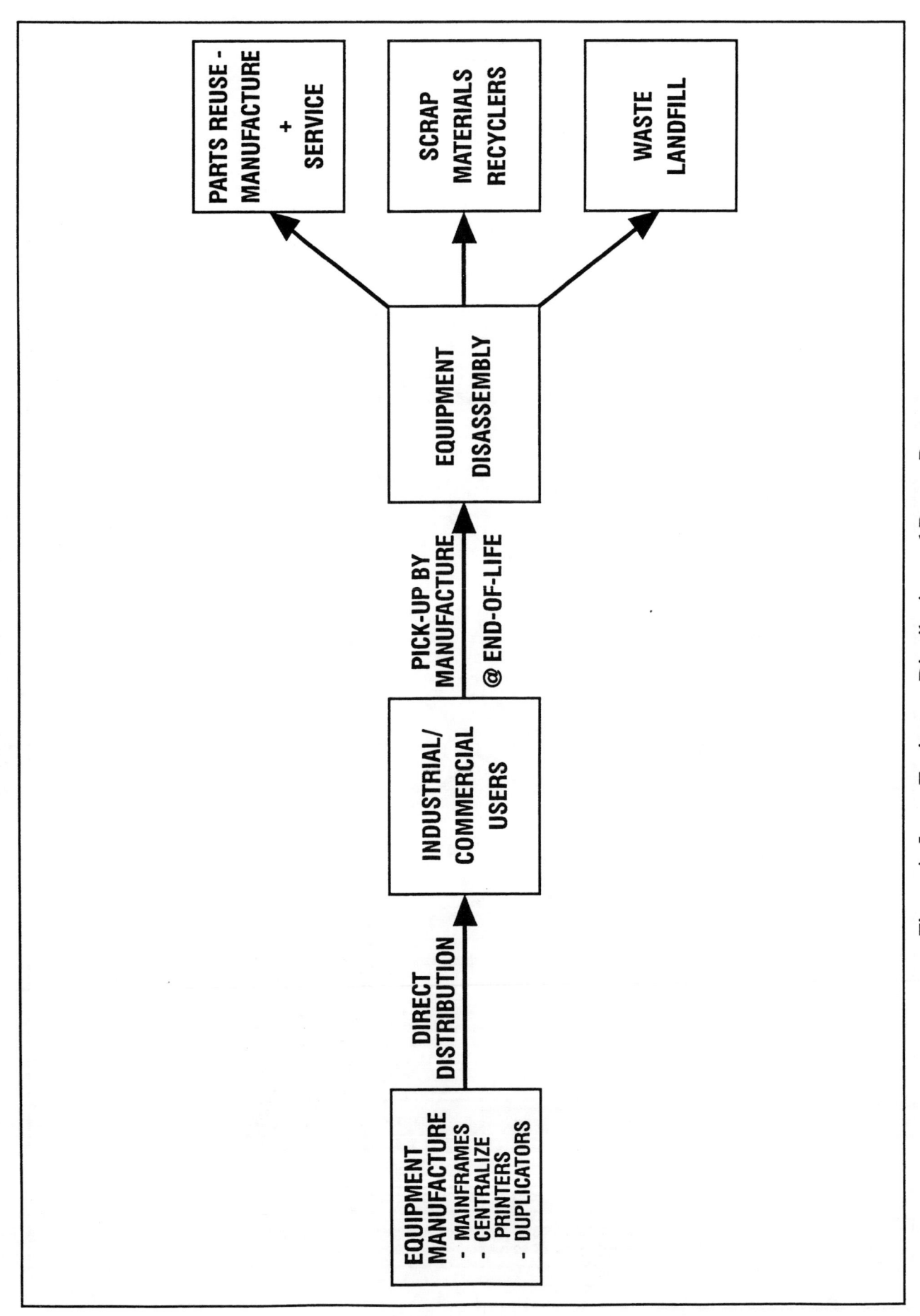

Figure 4. Large Equipment Distribution and Return Process

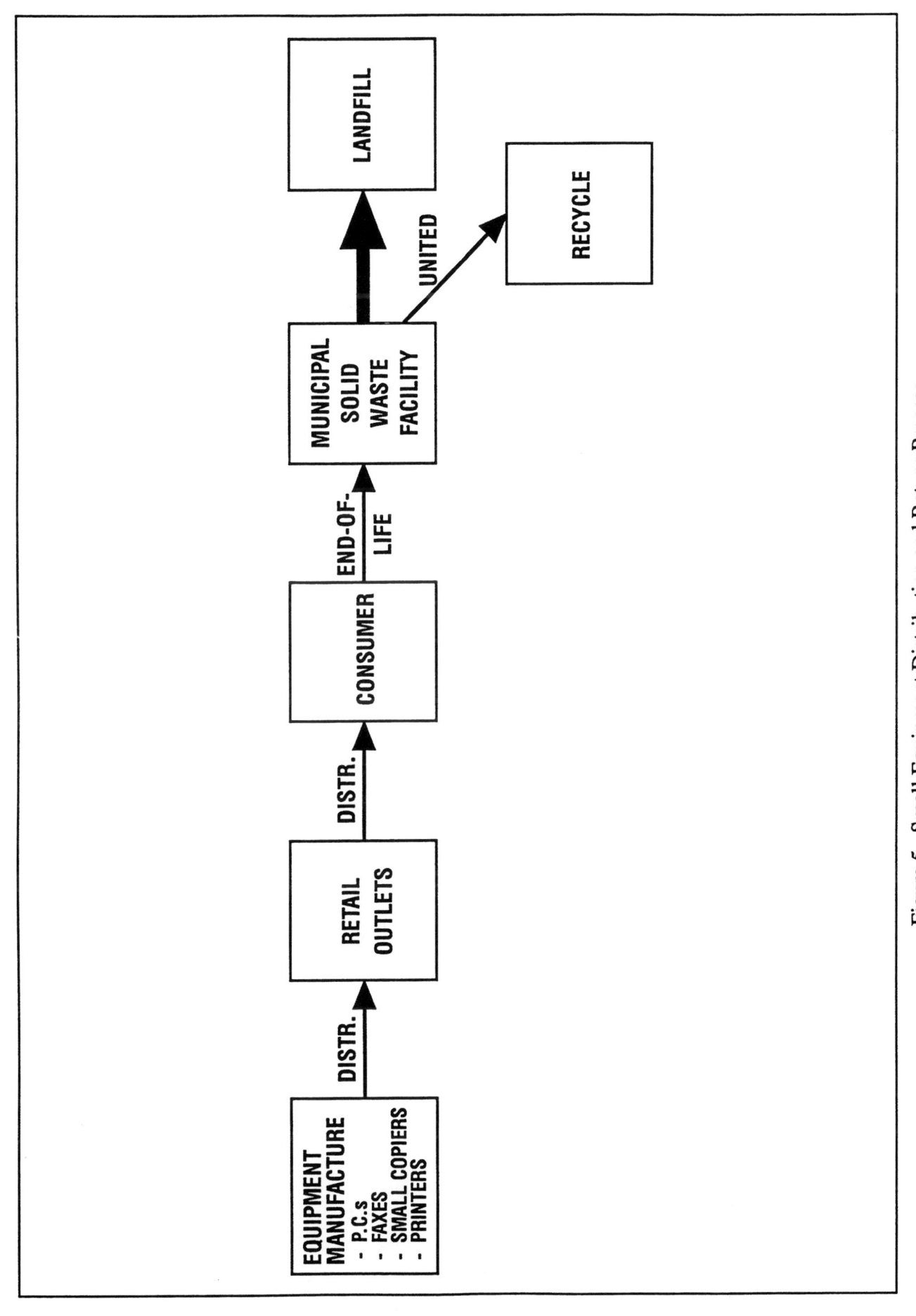

Figure 5. Small Equipment Distribution and Return Process

Approaches to *protect customer and OEM sensitive/confidential information*. Most OEMs will not have enough volume of end-of-life products to sustain independent recyclers or to justify the cost of establishing a full-service recycling and disposition center. Recyclers will have to rely on business from many OEMs to optimize productivity and remain cost effective. Security issues will become very important in such an arrangement for both OEMs and end users of such equipment. Often, electronic and business equipment will have parts or components that retain memory and, in some cases, have image memory from usage. Techniques to neutralize such memory effects will remove barriers to common usage.

3.3 Refurbished Parts Can Become Valuable Assets

3.3.1 Issues. Incorporating used parts back into new equipment requires highly reliable technologies that ensure refurbished parts meet the same quality standards as required for new parts. This is especially important in order to dispel the current negative public perception surrounding used parts and machines.

The open-loop alternative is to use these parts in other products/markets. In this case, less rigorous standards may apply permitting less extensive refurbishing and repair processes. This opportunity is discussed in Sec. 3.8 Improving the Economies Through Market Pull.3.3.2 Opportunities. Pollution prevention opportunities related to refurbished parts include the following:

> *Refurbishing technologies.* Most refurbishing processes include at least one cleaning operation. Current technologies still rely heavily on cleaning solvents such as halogenated organics and, more recently, semi-aqueous and aqueous materials.
>
> Emerging technologies tend to favor dry systems such as carbon dioxide pellet blasting [7]. Xerox Corporation is using this technology to clean machines and subassemblies in their remanufacturing process [8]. AT&T applies the same technology to clean fixtures used in manufacturing processes [9]. While technologically effective for more robust parts, in many cases, these techniques need further development before they are applied to more sensitive parts such as circuit boards and computer chips.
>
> For more sensitive electronic components, Texas Instruments [10] and AT&T [9] are evaluating the feasibility of supercritical carbon dioxide cleaning technology. It is anticipated that this technology will become viable for this application in the next two to three years.
>
> *Test procedures to ensure reliability and performance.* Some electronics and office equipment manufacturers are currently working to develop a technique that will aid in determining reused part life. This technique is known as Signature Analysis. Signature analysis is a method used to capture the key intrinsic characteristics of components through electronic signal conditioning techniques. These "signatures" will be used in the characterization of newly manufactured products as well as remanufactured and recovered parts. Life expectancy profiles can also be deduced to support component aging decisions.
>
> Xerox, for example, is applying this concept to solenoids [11]. An intrinsic signature waveform is constructed by plotting changes in current from the time the solenoid is energized until it reaches steady-state condition. Comparing

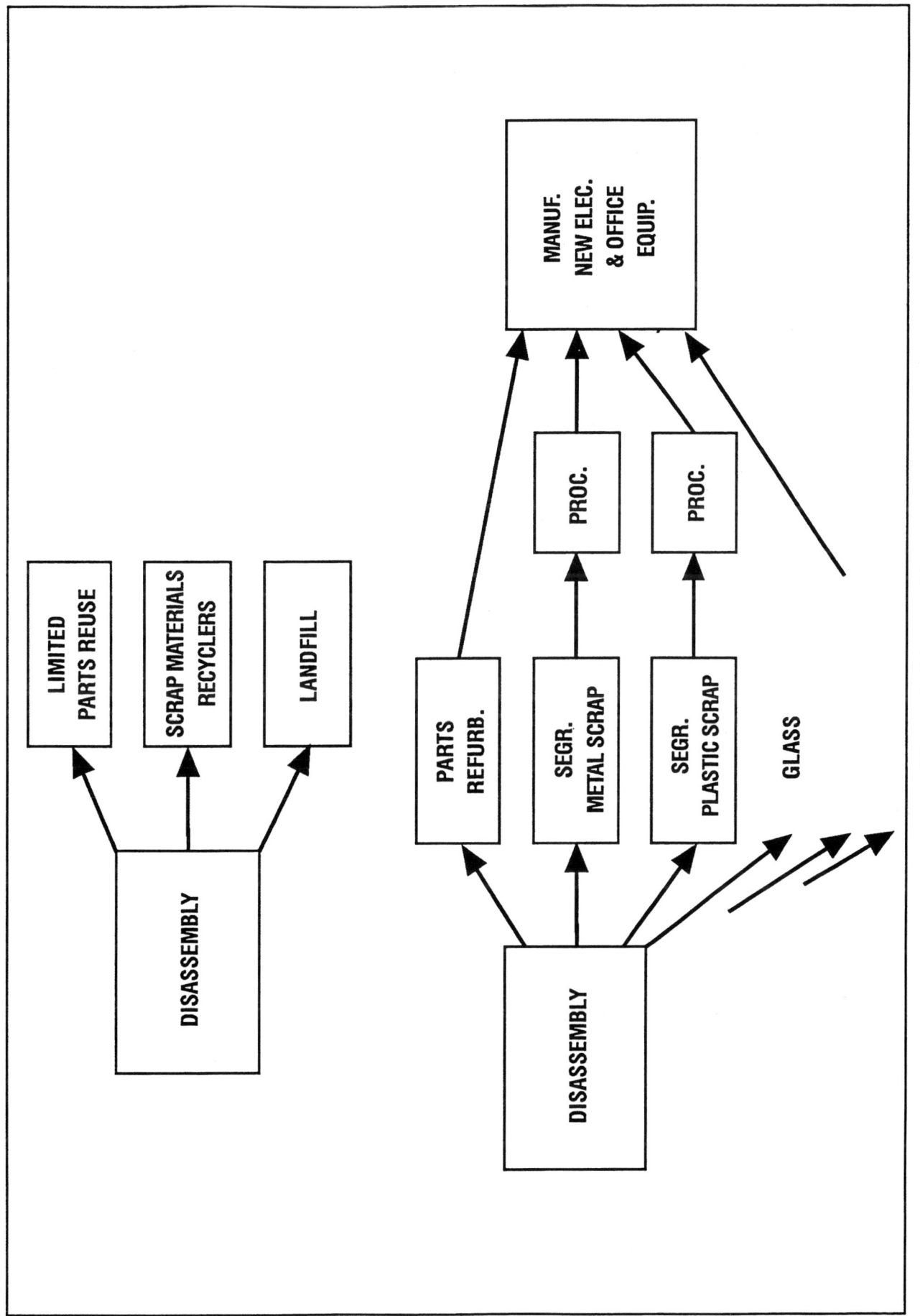

Figure 6. Disassembly

used solenoids against this control waveform will allow Xerox to make some determinations as to the expected future performance of the part. Additionally, an aging pattern is inferred from changes in this wave form over the life of the solenoid. By plotting this information, Xerox expects to be able to determine part age as well as life expectancy. With further development over the next couple of years, the company hopes to increasingly benefit from this technology as it becomes an integral part of remanufacture and service procedures.

Technologies for tracking part life and number of turnovers. In Europe, the EUREKA project focuses on technology for easily tracking part life [12]. As part of this project, Sony is developing a relatively low-cost memory chip for recording part history. This chip can serve as the mechanism for tracking the age and indirectly performance of individual parts.

3.4 Recycling Recovers Raw Material Value of Scrap

3.4.1 Issues. The majority of electronic and office equipment scrap materials fall into two general categories: (1) metals and (2) engineering plastics. The primary needs today centers on plastics. Since a recycling infrastructure for metals already exists, there is an economic incentive to recycle these materials. In contrast, the lack of an infrastructure and cost-effective technologies has made recycling of engineering plastics difficult. While the economic benefits have been demonstrated in isolated cases, much development is needed for wide-scale implementation.

Converting engineering plastics back into electronic and office equipment requires processing that includes separation, cleaning, and grinding. Companies that are currently recycling such materials use labor-intensive processes to separate and clean scrap plastics. To date, economics have limited volumes to large scrap plastic parts of known resin type with minimal contamination from other materials (e.g., metals, adhesives, labels, and other plastics).

Maximizing scrap plastics recovery will probably require automated separation and cleaning processes. Existing technologies are now being surveyed and evaluated at a pilot scale to demonstrate technical feasibility. While some off-the-shelf technologies have proven satisfactory, others need further development. Ultimately, the optimum combination of technologies must process high volumes of scrap at low cost.

3.4.2 Opportunities. Pollution prevention opportunities related to recycling to recover raw material are discussed below.

> *Plastics identification and sortation.* One impediment to closed-loop plastics recycling is cross contamination of differing plastic resins. In the past, office and electronic equipment parts were made from a wide variety of plastic resin formulations. Mixing resin types often degrades mechanical, thermal, and rheological properties such that the recycled resin cannot be re-processed into acceptable parts. Since plastic resin type is virtually impossible to determine through visual inspection, other means must be developed to effectively separate differing resin formulations. To complicate matters even more, these technologies must be able to accommodate surface treatments and additives such as paints, coatings, and pigmentation. Until adequate marking systems have been implemented (see Sec. 3.6), technology that includes rapid identification and separation is needed (see Sec. 3.2).

Contaminant/unwanted material removal. Technologies are needed to efficiently remove unwanted contaminants such as ferrous and non-ferrous metals, glass, rubber, paper labels, adhesives, foam, paint and metal coatings, etc. from plastics. The American Plastics Council is currently demonstrating the technical feasibility of combining various off-the-shelf separation and sortation processes that will effectively process scrap office equipment to clean plastic flake [13]. The electronic and office equipment industry expects to be able to utilize this type of processing technology on a commercial basis within the next two years.

Improving the *composition and properties of recycled plastics.* Recycled materials often have degraded properties when compared with virgin materials due to exposure to UV radiation and fatigue from extended use. Such materials will have to be reformulated and/or recompounded with additives and fillers to improve their mechanical properties. Alternatively, new resin technology may be used to modify resin structures and thereby improve properties.

3.5 Improving Recycling Through Material Compatibility

3.5.1. Issues. A major impediment to materials recycling is the complex process of separating out incompatible materials. For example, different plastic resins need to be separated from one another before being recycled. Without this separation, the newly molded parts will not meet critical property requirements. Making materials more compatible may eliminate the need, and associated cost, of separating some materials prior to recycling while enabling recycled resin blends with adequate properties.

3.5.2. Opportunities. Opportunities to prevent pollution through improved recycling via material compatibility include:

> *Increased compatibility between different materials* will eliminate/reduce the need for separation. As shown in Fig. 7, to date, some plastics compatibilities have been evaluated. However, technology needs to be developed to increase the variety/type of materials that can be recycled without separation. With appropriate modification and additive technologies, resin suppliers are ideally suited to extend and enhance resin system compatibilities within the next two to five years.

3.6 Designing In Product End-of-Life Value

3.6.1. Issues. The learning process for future design is largely based on current efforts to maximize the value of equipment reaching end-of-life. Office and electronic equipment manufacturers have drawn on this experience to identify important design features that will enable improved end-of-life management practices. How these features are considered in the product design phase depends on the desired objective(s). Will the product at end-of-life be reused, remanufactured, or recycled? Additionally, other design objectives (e.g., ease of assembly, service, etc.) as shown in Fig. 8, cannot be sacrificed in the process of improving end-of-life design.

Xerox has been successful at incorporating these principles into some office equipment parts and assemblies. Most notable to date, is the redesign of copy cartridges used in low-volume copiers [14]. In old designs, assembly housings that contained the main xerographic elements were assembled using ultrasonic welding. While several main internal

elements could be reclaimed, the plastic housing had to be destroyed in the process. In the new design,
ultrasonic welds have been replaced with a few fasteners. Xerox can now totally remanufacture these assemblies and recovers over 95 percent of parts and materials.

Applying these same concepts to entire systems will require new innovative technologies. Criteria that need to be addressed include life extension, easy disassembly, reuse, and recyclability.

3.6.2. Opportunities. Opportunities to prevent pollution by designing in product end-of-life value include:

> *Attachment technologies/techniques* (e.g., alternatives to metal inserts)
>
> - easily separable
> - sturdy & durable
> - able to be reused multiple times
> - compatible for materials recycling
>
> *Marking plastic parts for easy identification*
>
> - rapidly decipherable
> - read automatically (e.g., optical technology)
>
> *Label/marking technologies*
>
> - removable through non-destructive and simple techniques (applicable to parts being reused & remanufactured)
> - compatible with materials recycling won't contaminate recycle stream
>
> *Aesthetic quality technologies*
>
> - surface finish technologies eliminate need for decorative paint on structural foam parts; durable for multiple reuse
> - pigmentation/color technologies achieve color matching that is compatible with materials recycling

3.7. Design Tools

3.7.1. Issues. Designing for disassembly and recycling adds to an already complex set of requirements that design engineers must meet. Proper tools have become an essential means for designers to satisfactorily meet a concurrent array of requirements such as assembly, manufacturability, serviceability, and disassembly without sacrificing cost and time-to-market. Companies such as IBM and Xerox are currently integrating environmental dimensions into their product design processes that will require innovative tools for success [5,6].

3.7.2. Opportunities. Opportunities for pollution prevention through designing for disassembly and recycling include:

> Software tools for *rapidly assessing the quality of a design for disassembly*, evaluates potential impact on other design criteria (e.g., assembly, service), and aids in selecting lowest-cost alternatives. Carnegie Mellon University has focused research efforts on studying cost benefit aspects of designing for recyclability [15].

	Lexan PC	Cycolac ABB	Noryl PPO/PS	Noryl GTX PPO/PS Nylon	Valox PBT	Heavy Valox PBT/Ceramic	Xenoy PBT/PO	Lomod PBT/Poly THF	Geloy ASA	Ultem PEI	Cycoloy ABS/PS	Poly-styrene	Cryst-aline Nylon
Lexan PC	-												
10101010 Cycolac ABB	2	-											
Noryl PPO/PS	1	0	-										
Noryl GTX PPO/PS Nylon	0	0	2	-									
Valox PBT	2	2	0	0	-								
Heavy Valox PBT/Ceramic	2	2	0	0	2	-							
Xenoy PBT/PO	2	2	0	0	2	2	-						
Lomod PBT/Poly THF	2	2	1	1	2	2	1	-					
Geloy ASA	2	0	0	0	1	1	1	1	-				
Ultem PEI	1	0	0	0	0	0	0	0	0	-			
Cycoloy ABS/PS	2	2	0	0	2	2	2	2	1	0	-		
Polystyrene	0	0	2	1	0	0	0	0	0	0	0	-	
Crystaline Nylon	0	1	1	1	0	0	0	0	0	0	0	0	-

2 Compatible
1 Compatible to be a certain level
0 Not compatible

Figure 7. Material Selection Compatibility of Polymers

DESIGN FEATURES	AFFECTED ACTIVITY				
	Design for Assembly	Reuse	Reman	Recycle	Service
		Asset Recovery Management			
Part consolidation	-	X	X	X	X
Design parts to be multifunctional	-	X	X	X	X
Use standard components and processes	X	X	X	-	-
Develop a modular design approach	X	X	X	X	X
Minimize orientation of parts	X	-	-	-	X
Facilitate insection/alignment	X	X	X	X	X
Reduce fasteners, springs, pulleys, harnesses	-	X	X	X	-
Minimize adjustments	-	X	X	-	X
Design for commonality	X	X	X	X	X
Utilize similar materials	X	X	X	X	X
Minimize use of adhesives	-	X	X	X	X
Utilize marking codes	-	X	X	X	X
Utilize molded-in color	X	X	X	X	X

X Applies in all cases - Does not apply

Figure 8. (3Rs) Design Principles Matrix

Methodology to recommend best product/system architecture to enable *simplified retrieval of hazardous wastes and highest value parts*.
Software technology that recommends *most appropriate materials for a design function* considering recyclability and reuse with other common parameters including cost, performance, and ease of manufacturing.

3.8 Improving the Economics Through Market Pull

3.8.1. Issues. Remanufactured products and, to a lesser extent, products with recycled materials content may be viewed by the public as lower quality, undesirable products. Some governments perceive remanufactured/recycled goods as "used" and restrict purchasing of such products. As a result, markets for recycled materials and remanufactured products have been slow to develop and prices, due to low volumes, remain artificially high.

3.8.2. Opportunities. Opportunities to improve economics through market pull include:

Alternative applications and new markets for end-of-life products and used components. Often secondary markets can be simpler and faster to develop than original applications in primary markets.
New products containing high quality recycled materials. Market data indicate that customer acceptance of products with quality recycled materials can be high.
Improve perception of "used/remanufactured" parts and equipment (e.g., create incentive systems, education, and awareness). Education and awareness are important to build customer confidence as well as economic incentives to promote purchase.
Revise state and local acquisition/procurement policies. Many state and local procurement policies currently prohibit or restrict the purchase of used/remanufactured office and electronic equipment. Quality products and factory remanufacturing (as opposed to reconditioning) can overcome such objections.

3.9. Other Issues

Potential *anti-trust* barriers. In order for OEMs to work together on industry-wide problems, potential anti-trust issues need to be resolved proactively.
Protecting product technologies from third party use. Introducing products that can be recycled makes it easier for recyclers and remanufacturers to compete with OEMs. Reverse engineering becomes easier and, therefore, is more likely to occur.
Unauthorized recovery/sale of products. Another consequence of developing products for remanufacturing and recycling is unauthorized recovery and resale. OEMs can overcome this to some extent with software strategies that are proprietary.
Safety Agency Certification. To receive UL certification for plastics, current requirements mandate material traceability. While this may be appropriate for virgin resins originating from petroleum/chemical feedstocks, it presents a significant obstacle to using recycled plastics particularly if they are collected from multiple sources. A more acceptable approach would be to base certification on functional testing requirements, independent of sourcing.

4. References

[1] J. P. Desgeorges, Study Report on Valorization of Electrical and Electronic Products, Paris, December 4, 1992.

[2] Dispostion Breakout Session Report, Microelectronics and Computer Technology Corporation Electronics Industry Environmental Roadmap Workshop, November 10, 1993.

[3] H. Frankel, Center for Plastics Recycling Research, the State University of New Jersey, Personal Communication, 1992.

[4] M. Biddle, Technology Update - Durables Technical Project M-129, American Plastics Council Computer and Business Equipment Committee Meeting, Washington, DC, February 3, 1994.

[5] J. R. Kirby and I. Wadehra, Designing Business Machines for Disassembly and Recycling, Conference Record of the 1993 IEEE International Symposium on Electronics and the Environment, p. 32-36, May 10-12, 1993, Arlington, Virginia.

[6] V. Berko-Boateng, J. Azar, E. deJong, and G. A. Yander, Asset Recycle Management - A Total Approach to Product Design for the Environment, Conference Record of the 1993 IEEE International Symposium on Electronics and the Environment, p. 19-31, May 10-12, 1993, Arlington, Virginia.

[7] Environmental Alternatives, CO_2 Blast Cleaning of Disassembled Xerox Machines, April 1991.

[8] C. Genca, CO2 Blast Cleaning of Disassembled Xerox Machines, Xerox Disclosure Journal, 1991.

[9] J. Evans, U. Ray and P. Read, A Super Critical and Solid Case for Carbon Dioxide in Cleaning, AT&T Centerline, Bell Laboratories Engineering Research Center, Princeton, NJ, May 1993, p2-3.

[10] R. P. Lizotte, Jr. and D Weber, Supercritical Fluid Extraction (SFE) Cleaner Application, Conference Record Supplement of the 1993 IEEE International Symposium on Electronics and the Environment, May 10-12, 1993, Arlington, Virginia.

[11] C. R. M. Bartholomew, Signature Analysis: Strategic Weapon Against Time, Xerox Internal Report, December 9, 1993.

[12] Proceedings of the PREPARE Workshop on Cleaner Production and Cleaner Product Design for Electronic Consumer Goods, March 29-30, 1993, Stuttgart.

[13] M. Biddle, Technology Update - Durables Technical Projects M-130, M-131, and M-133, American Plastics Council Computer and Business Equipment Committee Meeting, Washington, DC, February 3, 1994.

[14] J. Azar, Asset Recycling at Xerox, EPA Journal, United States Environmental Protection Agency, Volume 19, Number 3, p. 14-16, July-September 1993.

[15] R. W. Chen, D. Navin-Chandra, and F. B. Prinz, Product Design for recyclability: A Cost Benefit Analysis Model and Its Application, Conference Record of the 1993 IEEE International Symposium on Electronics and the Environment, p. 178-183, May 10-12, 1993, Arlington, Virginia.

About the Authors

Jack C. Azar is the Manager of Environmental Design and Resource Conservation at Xerox Corporation. Patricia A. Calkins is the Corporate Manager of Resource Conservation projects at Xerox Corporation.

Pollution Prevention in the Metals Coating (Painting) Industry
Marvin M. Floer
Paint Systems Legal Requirements
Chrysler Corporation
Auburn Hills, Michigan 48236
(313) 576-1599

This monograph focuses on the automotive and light-duty truck coating industry to provide a representative example of the coating industry in general. Because customer expectations of an automotive paint finish are the most demanding in terms of appearance, quality, and durability, technological solutions for automotive finishing can likely be applied to less complex and less demanding coating industries. In practice, this is evidenced by the waterborne and powder coating systems already in operation for small parts and appliance coating. The same systems are still evolving or not yet feasible for automotive coating. Various combinations of the elements the materials, equipment, processes, principles and issues will retain their applicability when interrelated to other coating industries. The automotive coating industry also has the greatest need for assistance because environmental regulatory barriers are imminent and more restrictive.

Key Words
Add-on controls; air toxics; electrostatic application; end-of-pipe controls; cascading techniques; lowest achievable emission rate (LAER); metal conversion; photochemical reaction; powder coating; precursor; reasonable available control technology (RACT); transfer efficiency; volatile organic compound (VOC); waste stream; waterborne coating.

1. Background on the Automotive Coating Industry

The development of automotive coating technology over the last 15 years can be described as a dynamic process, driven by customer requirements, competitive pressures, and environmental regulations. Prior to 1978, coating technology had remained relatively static.

During the late 1970s, the U.S. automotive industry was suffering from both domestic and foreign pressures. Business objectives of automotive finishing could not be met with the traditional coating materials and facilities. The pattern was clear, since other industries, traditionally considered as American technologies, had disappeared from North America. Rapid and drastic change was clearly on the agenda of the domestic car manufacturers.

1.1. Customer Requirements. The purpose of any "paint job" is to provide an attractive, protective finish to the product. However, achievement of this goal is more complex than ever before in history. Any durable product in today's marketplace must meet the quality standards set by the customers. Blueprints, specifications, and performance standards must be consistent with these standards which will likely continue to increase and change many times during the life of any durable product. Today's manufacturers expend tremendous resources to acquire early customer feedback and to forecast consumer requirements, free market demand, and specific product preferences. A new product will lose its competitiveness if it is "late to market." Therefore, agility in manufacturing is critical to success in the automotive and most other industries.

1.1.1 Coating Appearance. A key factor in a prospective new car buyer's impression of the vehicles is the paint finish. A highly reflective, glamorous paint finish can captivate prospective customers from long distances. A discriminating car buyer will be attracted towards a paint finish that has high gloss levels, reflects a true image, has depth of image, and is free of imperfections such as dirt or mars. As a result, paint finish quality is sacred to a paint shop, and all supporting processes, equipment, and materials must be capable of meeting or exceeding quality objectives. The importance of the paint finish in swaying new car buyers is evidenced by the tremendous costs expended in paint facilities and coating materials. For example, until several years ago, the "paint job" was the most costly component of an automobile.

1.1.2 Protective Finish. A quality paint finish must provide the product with a durable coating that is resistant to harsh environmental extremes. More discerning customers and extended vehicle warranties in the last 15 years have demanded corresponding improvement in the performance of the coatings. Today's car buyers expect a paint finish that will hold up for a minimum of 10 years and provides:

> A durable, quality coating system with strong adhesion to the car body, as well as between several coating films
> A finish that retains its luster and resists ultraviolet fading over the useful life of the product
> A coating that is chemically resistant to atmospheric fallout (acid rain), biological secretions, gasoline, battery acid, road salts, etc.
> An impact absorbent coating that resists chipping from road gravel
> Corrosion protection that supports customer expectation.

1.2. Competitive Pressures. In the late 1970s, the Japanese automakers, with strong community, supplier, government, and financial institution support, had secured a solid 25 percent of the U.S. market. In contrast, U.S. manufacturers were prevented by Anti-Trust Law from any collaboration, even non-competitive product development, such as safety features, tail-pipe emission controls, and fuel economy. Japanese manufacturers in the U.S. seized opportunities to build new and modern facilities with low operating costs; high technology equipment, processes and materials; unprecedented tax incentives; non-bargaining labor; young employees; low health care costs; zero pension obligations; and "friendly" environmental regulatory locations. These foreign competitive advantages forced the domestic manufacturers to become more competitive in the areas of product quality, product development, and operating costs. All capital expenditures were directed towards new product development and facility improvements that added value to the product and the process. Thus, U.S. manufacturers did not have capital funds for discretionary or non-value added activities, such as end-of-pipe pollution controls.

1.3. Environmental Regulatory Initiatives. Since promulgation of the Clean Air Act in 1963, the U.S. Environmental Protection Agency (EPA) has recognized coating processes and specifically automobile and light-duty truck paint shops as major sources of VOC emissions, or evaporative hydrocarbons. Because VOCs and nitrogen oxides (NO$_x$) are precursors of low-level atmospheric ozone, the EPA has labeled them criteria pollutants. Subsequent amendments to the Clean Air Act sought further VOC emission reductions. The Clean Air Act Amendments (CAAA) of 1977 and associated regulations, guidelines, and policies had a technology-forcing effect on the automotive industry.

During the 1980s, air quality regulations forced the automotive industry to convert to costly high solids/low VOC basecoat-clearcoat paint finishes to meet environmental requirements. Conversion to basecoat-clearcoat coating systems forced automotive assembly plants to replace entire coating lines. Large capital reinvestment requirements led to closure of many older assembly plants when the facility did not warrant the large expenditure.

Similar environmental laws have been imposed only recently in Europe and are yet to be seen in Asia. Throughout the 1980s, overseas competition enjoyed a distinct cost and paint finish quality advantage through the continued use of low solids coatings with high solvent content. Low solids solventborne coatings provide a paint finish that cannot be achieved with high solids coatings.

Low solids coatings are also more "forgiving" in the application process. High solids coatings narrow the application window and reduce the gloss levels and distinctness of the reflected image, (i.e., the "orangepeel" phenomenon). (A basecoat/clearcoat finish does provide superior depth of image characteristics and superior coating durability.) The necessary change from low solid/high solvent to high solid/low solvent required U.S. manufacturers to tighten process controls to optimize the paint appearance. Higher coating material costs resulted in application efficiencies to reduce coating usage. Paint shops that utilized process controls and improved coating efficiencies were able to dramatically reduce the cost and quality disadvantages. Unfortunately, not all paint shops were successful.

The cathodic electro-deposition primer coating process is an excellent example of a pollution prevention success story that provided benefits to both the industry and air quality. This process employs a low VOC waterborne material that is dip-applied with nearly 100 percent efficiency and generates minimal waste. Most of the coating VOCs are released in the curing oven, and treated by an end-of-pipe afterburner making the add-on control efficient and cost effective (due to the low volumes of air exhaust containing high concentrations of VOC). Since the process is more efficient and offers increased corrosion protection, it has become the process of choice in the industry. However, it also required very significant financial reinvestment during a time when capital was scarce and in high demand for competitive product development.

1.4. Perspective on Pollution Prevention. Background on the (automotive) coating industry is intended to provide the reader with perspectives on the needs and directions of the industry and why the pollution prevention approach is preferred by industry over end-of-pipe controls. Background on the automotive coating industry is necessary to understand:

"Pollution Prevention Pays."[1] Reduction of chemical releases to the environment must be achieved within normal business objective constraints. Freedom from environmental regulatory pressures, cost savings, processing efficiencies, waste/scrap reductions and quality/product improvements are examples of routine corporate goals that are attracted to adequately planned and developed pollution prevention technologies. The term "environomic" has recently become popular in describing processes, equipment and materials that are cost-effective and environmentally beneficial.

In September 1993, the American Automobile Manufacturers Association reported to the EPA and Department of Commerce that *in the 1970s, paint finishing of each compact car released between 18.5 and 50 pounds of VOC emissions* to the atmosphere (Fig. 1 and Fig. 2).

- The industry complied with the CAAA of 1977 early in the 1980s, costing an average of $18 million per facility to reduce the emissions to 8.7 pounds per unit built. This met the EPA designation of RACT compliance. The VOC reduction (1,250 tons/year) came at a price tag of $14,400 per ton of VOC controlled.
- In the mid 1980s, the cost went to $16,000 per ton VOC reduction (188 tons/year at 7.2 pounds VOC/vehicle). The EPA designates this level of compliance as LAER which is determined by an examination of the lowest emitting technologies.
- For a newly constructed or modified paint shop of the 1990s, "state-of-the-art" environmental controls will cost $62 million, and will reduce emissions to 3.1 pounds per compact vehicle. This new LAER compliance level represents technology advances beyond those available in the mid-80's. Current control costs have risen to $80,000 per ton of VOC removed. This cost escalation is due to the higher costs to install and operate spray booth carbon adsorption and/or incineration.
- Continued reliance on end-of-pipe controls to achieve environmental improvement is not cost effective. End-of-pipe controls typically consume large amounts of energy and add pollution of greenhouse gases. *Add-on controls cannot add value to the product or process.*

Relaxation of Anti-Trust Law, the CAAA of 1990, the current executive administration, and the new EPA Administrator are examples of governmental cooperation with industry to reduce environmental releases by economically

---

[1] The terms "Pollution Prevention Pays" and "3P" are the property of 3M Corporation and are protected by copyrights.

VOC EMISSION PROFILE FOR A TYPICAL PAINT SYSTEM
(Pounds Per U.S. Compact Car)

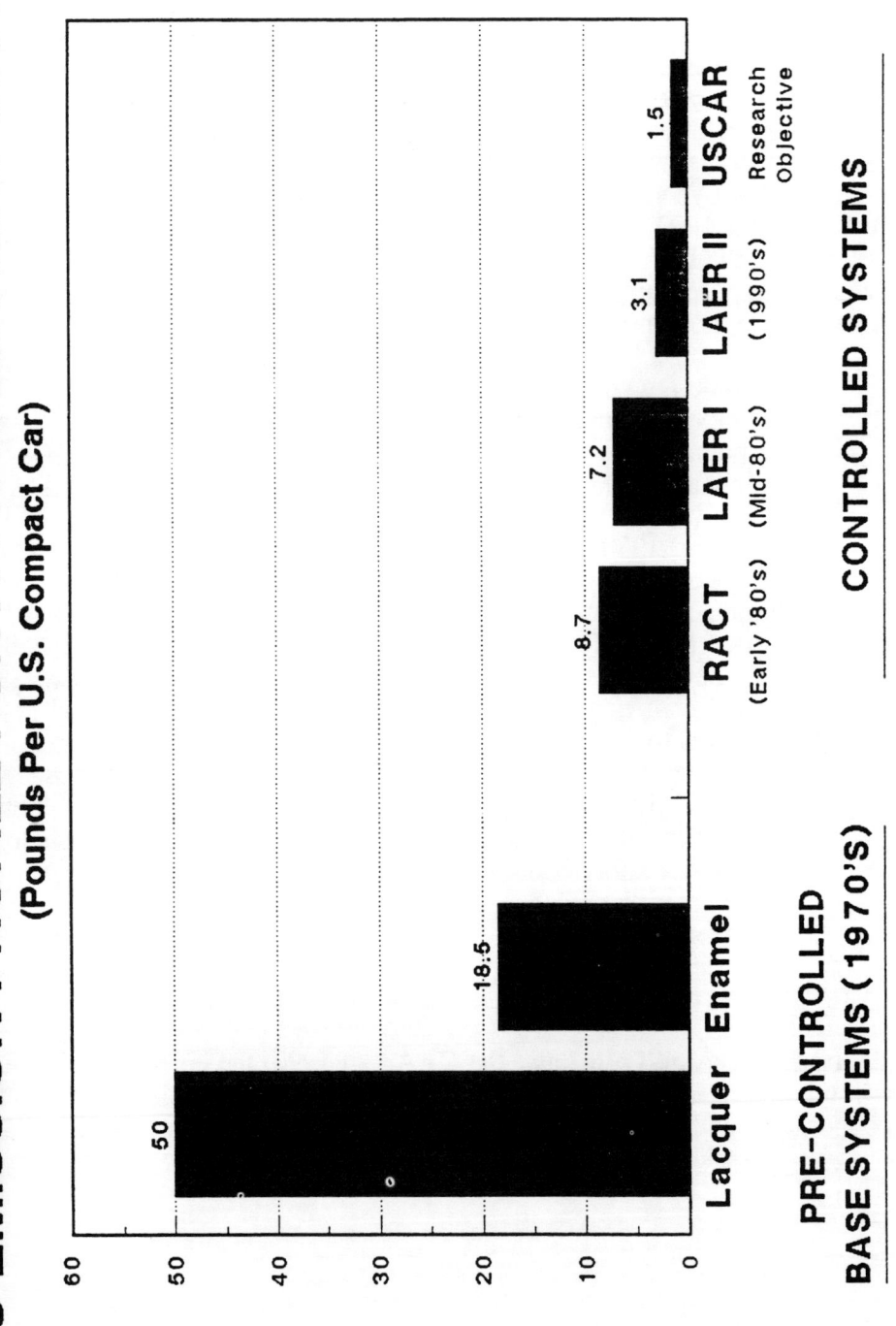

Lacquer	50
Enamel	18.5
RACT (Early '80's)	8.7
LAER I (Mid-80's)	7.2
LAER II (1990's)	3.1
USCAR (Research Objective)	1.5

PRE-CONTROLLED BASE SYSTEMS (1970'S) — CONTROLLED SYSTEMS

Figure 1 – Courtesy of American Automobile Manufacturers Association

VOC EMISSIONS VS. COST PER INCREMENTAL TON REMOVED
(Incurred Annualized Dollars)

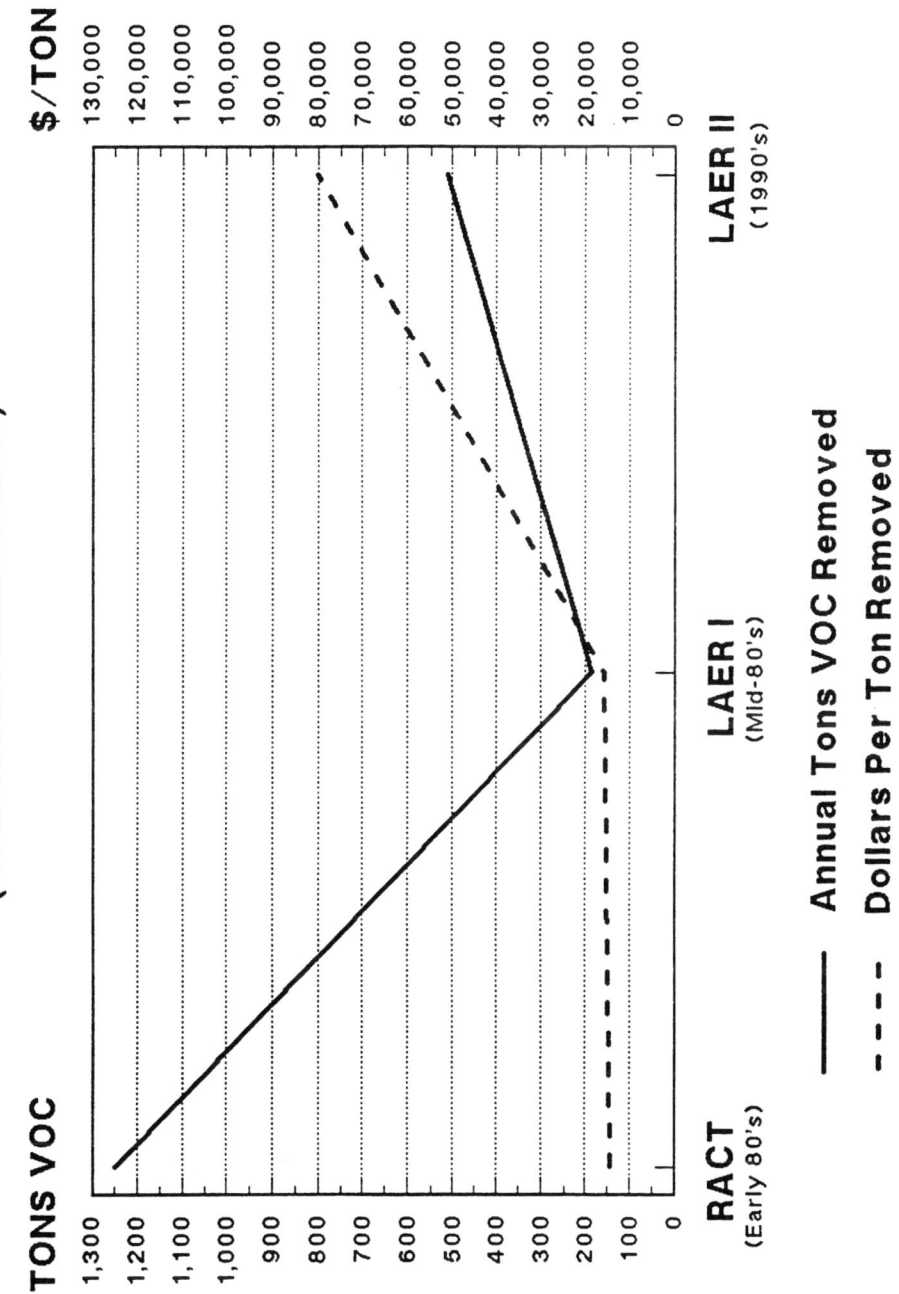

Figure 2 – Courtesy of American Automobile Manufacturers Association

sensible means. Sustainment of this display of confidence must be supported with significant, genuine results from industry.

In 1993, the United States Council for Automotive Research (USCAR), Chrysler Corporation, Ford Motor Corporation, and General Motors Corporation formed the Low Emissions Paint Consortium (LEPC). The LEPC will test and develop automotive coating application technologies to reduce coating emissions of a compact vehicle to *1.5 pounds VOC/vehicle without any end-of-pipe controls. This effort will also improve paint finish appearance and durability.* Technology feasibility is anticipated at the turn of the century. Unlike end-of-pipe controls, *Pollution Prevention means added value* to the product or process.

In a June 15, 1993 memorandum on "Pollution Prevention Policy Statement: New Directions for Environmental Protection," EPA Administrator Carol M. Browner ".... establishes a bold national objective for environmental protection: '[T]hat pollution should be prevented or reduced at the source whenever feasible.' .... We have learned that traditional end-of-pipe approaches not only can be expensive and less than fully effective, but sometimes transfer pollution from one medium to another. Additional improvements to environmental quality will require us to move upstream to prevent pollution from occurring in the first place."

2. Current Automotive Coating System Technologies

This section describes a typical composite automotive paint shop processing that utilizes prevailing coating material application processes. Due to the dynamic evolution of coating technology (Fig. 3 and Fig. 4), it is doubtful that all of the emerging technologies will exist in any one paint shop currently in operation.

The most common coating system (Fig. 5) consists of five key steps: (1) a metal cleaning and phosphate pretreatment conversion of galvanized steel substrates, (2) a cathodic electrodeposition primer film, (3) a coating of chip-resistant primer on critical surfaces, (4) a film of primer surfacer on all exterior surfaces, and (5) finally, a topcoat consisting of a base color coat and clearcoat film. All spray coatings discussed in this section will be addressed as solventborne coatings. While major coatings are not in common use in waterborne technologies, waterborne materials have been in place for some less critical coatings (i.e., interior small parts, blackout, chassis black, and underbody deadener).

Automotive coating technology has been in a state of transition for several years and will likely continue to evolve throughout 2000. This evolution is a direct result of the influence of customer requirements, competitive pressure, and environmental requirements.

2.1. Phosphate Pretreatment. Phosphate pretreatment, the first step in the coating process, consists of varying combinations of spray and/or immersion treatment stages to clean, chemically treat (conversion), and rinse the vehicle body. Three to thirteen stages can be used to form a layer of phosphate and/or manganese phosphate crystals on the metallic substrate surface. The resulting phosphate crystal conversion layer:

> Provides additional *corrosion protection*.

> Provides excellent *adhesion with paint*. The phosphate film is porous and absorbs the liquid paint. After curing, the phosphate film mechanically locks between the two coatings.

AUTOMOTIVE ENAMEL TOPCOAT TECHNOLOGY LIFE CYCLE

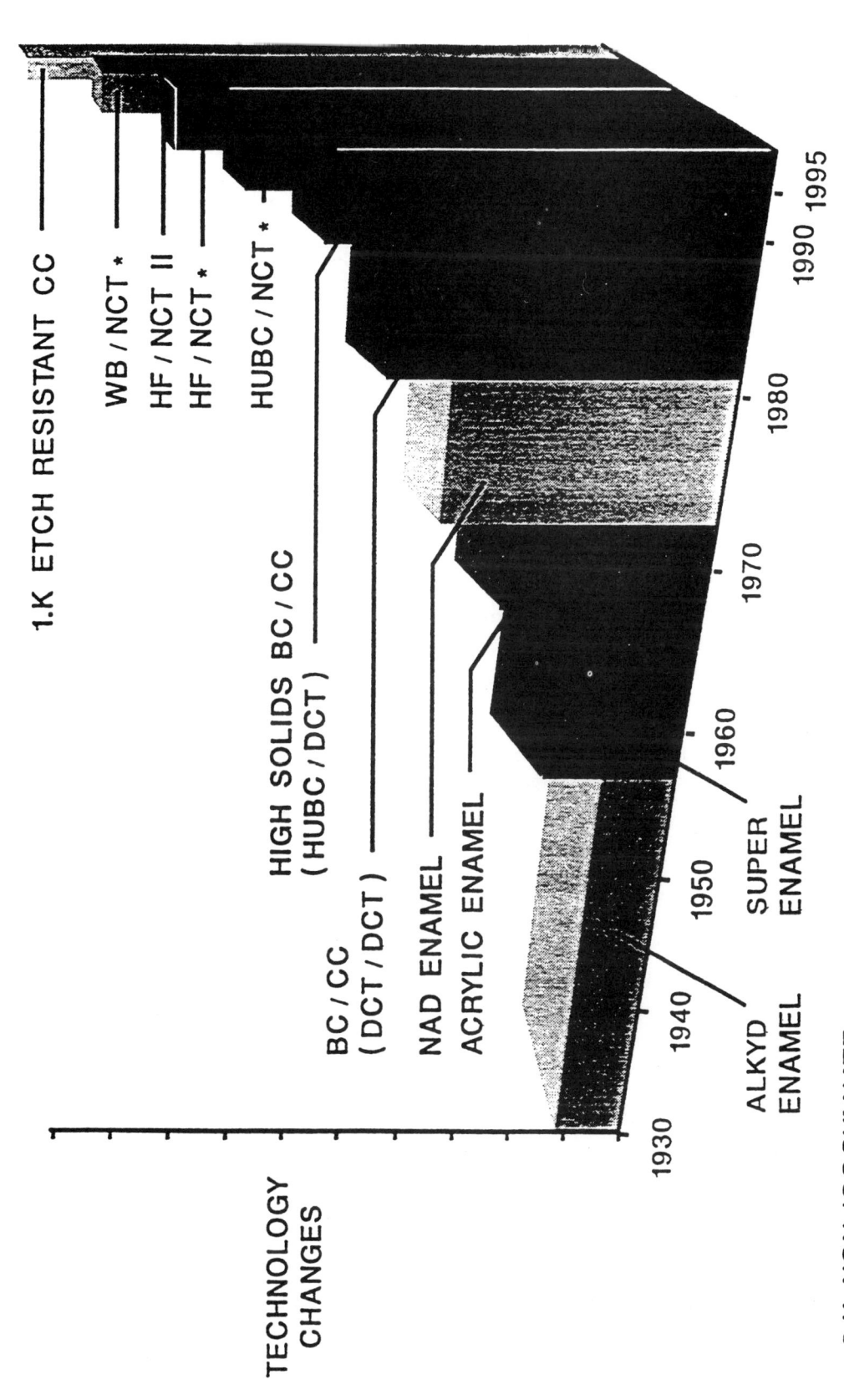

TECHNOLOGY CHANGES

- ALKYD ENAMEL
- SUPER ENAMEL
- ACRYLIC ENAMEL
- NAD ENAMEL
- BC / CC (DCT / DCT)
- HIGH SOLIDS BC / CC (HUBC / DCT)
- HUBC / NCT *
- HF / NCT *
- HF / NCT II
- WB / NCT *
- 1.K ETCH RESISTANT CC

1930 1940 1950 1960 1970 1980 1990 1995

* 2.K NON-ISOCYANATE

NAD = Non-Aqueous Dispersion
BC = Basecoat
CC = Clearcoat
HUBC = High Solids Universal Basecoat
DCT = Direct Cross-Linking Technology (Clearcoat)
NCT = New Coating Technology (Clearcoat)
2K = Two Part (Clearcoat)
HF = High Flow (Basecoat)
WB = Waterborne (Basecoat)

Figure 3 – Courtesy of PPG Industries

APPLICATION EQUIPMENT / PROCESS LIFE CYCLE

CHANGE TO APPLICATION METHOD

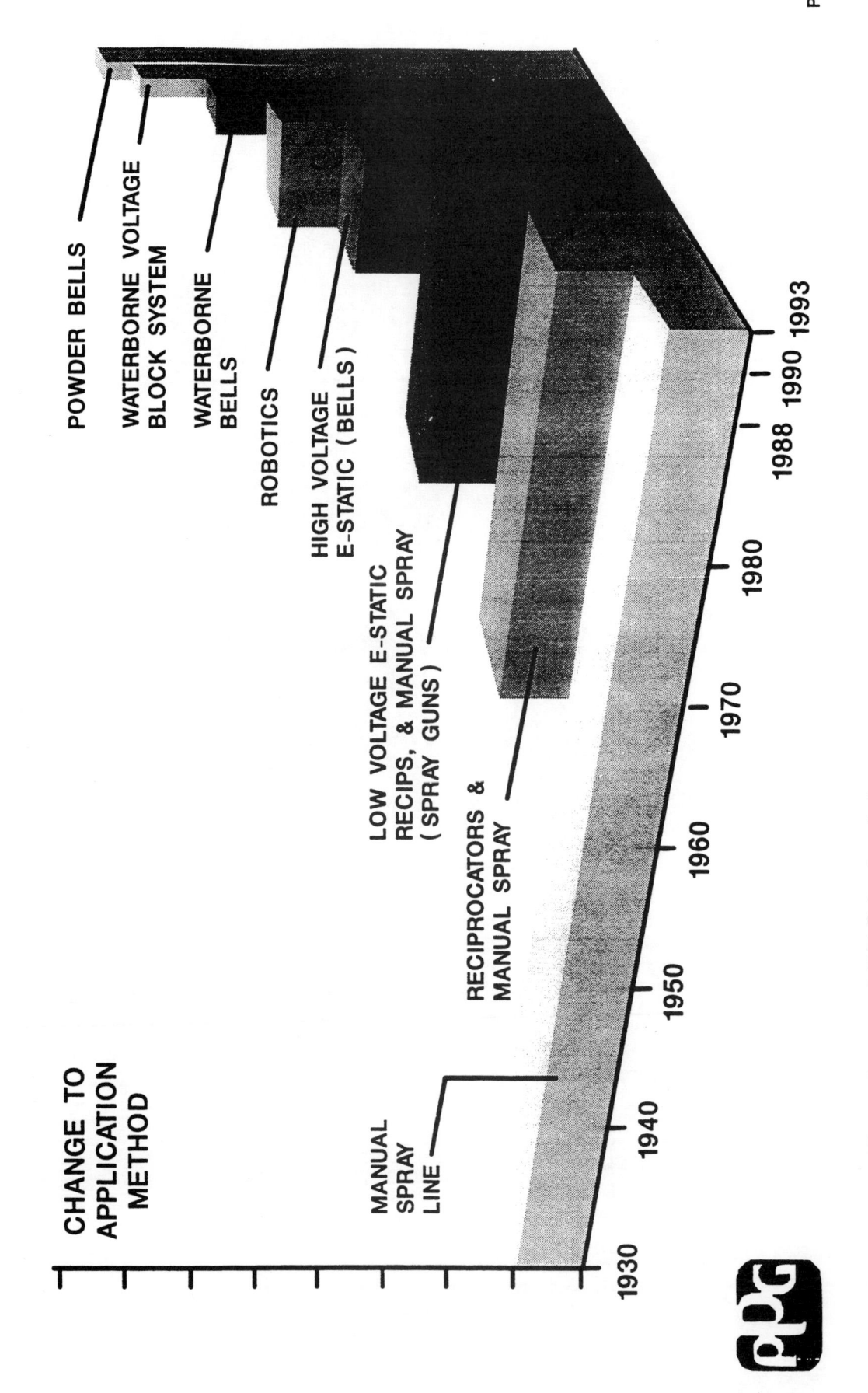

- MANUAL SPRAY LINE
- RECIPROCATORS & MANUAL SPRAY
- LOW VOLTAGE E-STATIC RECIPS, & MANUAL SPRAY (SPRAY GUNS)
- HIGH VOLTAGE E-STATIC (BELLS)
- ROBOTICS
- WATERBORNE BELLS
- WATERBORNE VOLTAGE BLOCK SYSTEM
- POWDER BELLS

1930 1940 1950 1960 1970 1980 1988 1990 1993

Figure 4 – Courtesy of PPG Industries

CLEARCOAT	1.8 - 2.5 MILS
BASECOAT COLORCOAT	0.6 - 1.0 MILS
PRIMER-SURFACER	1.0 - 1.5 MILS
ANTI-CHIP PROTECTION	1.5 - 4.0 MILS
ELECTROCOAT PRIMER	0.8 - 1.2 MILS
PHOSPHATE CONVERSION COATING	
TWO-SIDE GALVANIZED SHEET METAL	30 - 35 MILS

1 MIL = 0.001 INCH

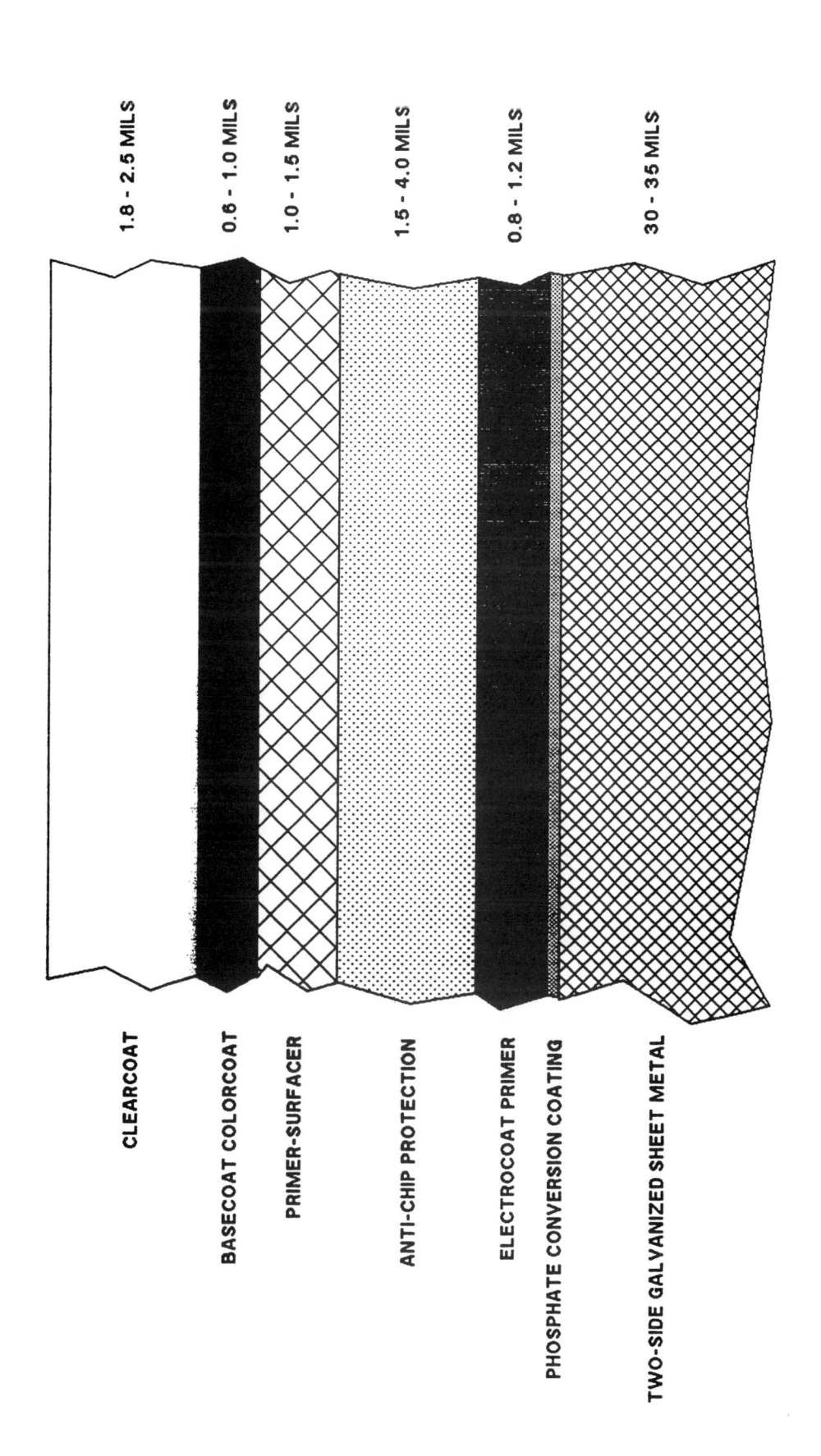

Figure 5 – Typical Automotive Coating System Currently in Use

> Offers *thermal expansion properties* that are intermediate between metals and paint, thereby smoothing out the thermal expansion differences.

In the 1980s, two key environmental improvements were made to this process: (1) conversion from hexavalent chrome rinse chemicals to the less hazardous trivalent chrome radical, and (2) reduction of phosphate composition with partial substitution of manganese-modified chemicals.

2.2. Electrodeposition Primer. Because of the marked advantages of electrodeposition, the process has become the dominant method for priming automobiles, as well as many other metallic products.

2.2.1. Advantages of Electrodeposition. A listing of the advantages of electrodeposition clearly explains the reasons for its spectacular success.

> Formation of protective films in highly recessed areas .... provides critical corrosion protection.
> Transfer efficiencies of better than 95 percent result in reduced paint usage and waste.
> Use of water as practically the only carrier virtually eliminates fire hazard, materially reduces water and air pollution, and markedly reduces (initial investment) cost of facilities for controlling these conditions.
> The low paint bath viscosity .... results in ease of pumping and allows drainage of the coated vehicle.
> Freshly-deposited paint is insoluble in water, permitting complete rinsing and recovery of dragged-out material.
> Unlike spray coatings, electrodeposited paint will not sag during baking.
> Unlike dip coatings, electrodeposited paint is not washed off in enclosed areas by hot vapors during curing.
> Since the process is automated, direct labor costs are markedly reduced.
> The deposited film is reproducible from part-to-part and day-to-day.

The electrochemical process of depositing paint appears, at first examination, quite complicated; however, it is actually more trouble-free than other paint application process for several reasons:

> The large tank volume tends to minimize variations in paint or process.
> The operating parameters are well defined and are based on extensive experience.
> Samples from the tank .... often predict problems before they appear on the line.

2.2.2. Advantages of Cathodic Electrodeposition. Although the principle of electrodeposition gained acceptance with the anodic systems, it was not until the cathodic method was developed that electrodeposition became the common method for priming automobiles. The cathodic method has the following advantages:

> It achieves corrosion resistance at low film thicknesses
> It permits throwpower to interior surfaces without penalty of overbuild on exterior surfaces

It achieves bimetallic joint corrosion resistance required by the increased use of galvanized metal

It possesses saponification resistance and long-term adhesion retention over an entirely galvanized panel

It requires substantially less electrical consumption for deposition and, as a result, less refrigeration for cooling

It achieves gloss and hold-out with and without spray primers

The primary advantage for cathodic systems, however, is superior corrosion protection. With anodic electrodeposition, metallic ions from the part being coated often become included in the deposited film. This condition results from anodic reaction when current is applied. The presence of iron ion in the film provides a readily available site for rust to start. In addition, when deposited, cationic resins are alkaline in nature and tend to be natural corrosion inhibitors.[2]

Cathodic electrodeposition primer material first became available commercially in 1975. It gained its popularity quickly.

2.2.3. The Cathodic Electrodeposition Process. Vehicle bodies are transported through the electrocoat process by an overhead conveyor with cradle-designed carriers. The conveyor declines, immersing the negatively charged body (cathode) into and through the electrodeposition tank containing the waterborne pigment paste and resin feed. Strategically located anodes, positively charge the immersion material and cause current to flow through the bath to the car bodies, beginning the electrodeposition of the paste/resin bath mixture onto all exposed areas of the car body. As the film thickness increases, its insulating properties assist in the control of the accumulation of the film thickness. Design and control of conveyor speed, tank length and electrical charge will produce the desired film thickness. Material formulation adjustments can also increase or decrease the material throwpower properties, providing high-film, medium-film and low-film material variations.

The conveyor incline removes the car body from the feed tank. Conveyor inclines and declines, as well as car body fill/drain holes, are designed to satisfy the necessary feed fill and drain requirements to maximize material contact, coverage and recovery.

Immediately after the electrodeposition process, the electrocoating is covered by a low density, porous film of non-deposited material solids, which must be removed prior to primer curing. The car body is processed through a full immersion rinse tank and then several spray rinse stages to remove the material. The rinses are staged in a reverse-cascading cycle with the dirtiest rinse first, progressing to a final virgin deionized water rinse. Each stage of the rinse cycle filters the rinse water and processes the rinsed material through an ultrafiltration system that recovers reusable feed material and returns it to the system. A 99 percent effective material usage can be achieved in a properly operated and maintained process.

2.3. Chip-Resistant Primer. The third step in the coating process involves application of chip-resistant primer coatings. These coatings are urethane-based materials, and are both resilient and impact absorbent to resist chipping from road gravel impingement. To achieve these properties, film thickness must be a minimum of 3 mils (0.003 inch). Historically, the primary coverage area has been the lower side sill areas where protection is most critical. In response to consumer demand, the coverage area has been expanded to include the leading-edge hood and front fenders. The customer's driving use and vehicle design may also impact

---

[2]Up to this point, Section 2.2 is derived entirely from "Automotive Electrocoat Reference Manual," PPG Industries, copyright 1989.

the coverage areas. For example, light-duty trucks and off-the-road vehicles are most likely to be exposed to road blast, while low, sleek design cars are more vulnerable to road chipping. When full-body coverage is desirable, chip-enhanced primer surfacer coatings can be used.

Until recently, chip-resistant coatings have been high viscosity, solventborne coatings sprayed with conventional air atomization. With difficulty, they can be sprayed electrostatically or with airless or semi-airless spray to increase the coating transfer efficiency and reduce material usage. High voltage electrostatic rotary atomizer application is the most efficient when the process and quality requirements can be met. Add-on emission controls are not practical on sill and/or leading edge processes because the booth air volumes are high and emissions available for capture are low. The resulting low concentration of emissions yields low control efficiencies.

PVC coatings also reduce underbody chipping due to their excellent protection, additional vibration damping, and road noise reduction properties. However, their high film thickness requirements (10 to 20 mils) leave a wavy substrate that cannot be used on critical appearance surfaces.

Waterborne chip-resistant primers are currently under development by coating suppliers. However, domestic automotive manufacturers agree that powder materials have surpassed results with waterborne primers. With the industry moving to a combined anti-chip/primer surfacer process, the zero-VOC powder will not require costly spray booth emission controls and will provide substantial improvements in customer satisfaction.

2.4. Primer Surfacer. Primer surfacer and chip-resistant coatings have a lot in common, but with the advent of chip-enhanced primer surfacer materials, the need for chip-resistant materials will soon be obsolete. Surfacer film thickness can be increased on critical surfaces to improve chip protection. The practical life of waterborne surfacer coatings is also expected to be short-lived with increasing attention given towards development of powder coating surfacer materials, their application, and associated equipment.

Primer surfacer, or guide coat, provides a film that lends "forgiveness" to the overall coating system. Surface coatings are used for a wide range of reasons:

> Added anti-chip protection
> To cover and fill surface imperfections
> Increased film flow for smoother surfaces to improve the distinctness of image of the color coat
> Strengthened intercoat adhesion
> Long term durability with excellent ultraviolet absorption properties that prevent color coat degradation which can lead to delamination.

Some manufacturers have successfully eliminated the primer surfacer process through the use of high film electrocoats. These thicker films allow some sanding of the surface. Elimination requires the manufacturer to:

> Process sheet metal free of major surface imperfections such as file or grind marks and drawing compound residues
> Control dirt in the electrocoat system and on the car body
> Increase ultraviolet absorption additives in the topcoats
> Exercise various improved process controls to prevent abnormal film conditions; for example, overbaked electrocoat.

2.5. Topcoat. The topcoat system most commonly in use in the U.S. is high solids, solventborne basecoat/clearcoat technology. The best finish is achieved with a thin basecoat film (0.6 to 1.0 mil) and a heavy clearcoat film (2.0 to 2.5 mil). Topcoat spray booths are predominantly designed with downdraft air flow in a manner that the paint particles are collected through baffled water wash curtains at very high efficiencies.

2.5.1. Basecoat. Basecoat materials provide the color aspect of a topcoat appearance. Color is determined by selection of pigments and metallic or mica-metallic flake additives.

After an exterior body solvent wipe and a tack-off, the initial spray of basecoat is done to interior coated surfaces (i.e., engine compartment, door jambs and inner door, luggage compartment) by either manual sprayers or robots. Spray guns are either conventional air atomized or low voltage electrostatic. Some more sophisticated coating systems may rely on a color-compatible, pigmented primer surfacer for the paint appearance in the closed compartments.

The first exterior spray zone builds most of the color film with either a set of nine or ten high voltage, high speed turbobells or a set of three reciprocating automatic machines. Very significant material savings and emission reductions are achieved through installation and optimization of turbobell applicators. They will apply about 75 percent of the paint solids sprayed. Reciprocators have a transfer efficiency about one-half of that and therefore must spray twice the paint to cover with the same film solids. Turbobells are capable of applying 100 percent of the basecoat film requirement with straight shaded colors (i.e., metallic-free coatings).

Metallic colors will lose brilliance and "go dark," compared to the styling master, when sprayed electrostatically. Therefore, metallic and mica-metallic colors require a second "dress coat" with reciprocators or robots to bring the color match back to the approved styling color. A conventional air atomized finesse coat is then applied manually on shy film areas or to "dress out" imperfections.

With each color change, paint lines and applicators must be purged of the spent paint. A typical basecoat spray booth zone contains 18 or more spray guns that must be purged. Waste material costs are high for purging. In most cases, the mixture of purging solvent and waste paint can be collected for remanufacture of purge solvent. Scheduling can minimize costs of purging by routing bodies of the same color in blocks or batches through the paint shop to eliminate many of the purges along with their associated waste and costs.

Drying ovens (usually infrared) are sometimes used to flashoff most of the basecoat VOC prior to clearcoat application. By "setting up" the basecoat film, a smoother finish can be achieved to improve appearance.

2.5.2. Clearcoat. Through full utilization of high speed rotary atomizers (turbobells), the clearcoat spray process has become a very efficient process for solventborne paints. Up to 85 percent clearcoat transfer efficiency can be achieved. Clearcoat requires only preventative purging.

The typical exterior clearcoat spray coverage consists of one or two sets of high voltage turbobells (nine or ten applicators per set). Although one set of turbobells is capable of covering at desired film thicknesses, paint will sag and limit the achievable film thickness, particularly on vertical surfaces. There are some processes that get by with one set of turbobells and others that will reinforce the verticals with an additional applicator on each side. Since the clearcoat film thickness determines the "depth of image," industry looks for ways to make increases and improve appearance.

Some interior surfaces are clearcoated, such as door jambs, with manual dressup spray. Hand-held electrostatic guns are very effective on clearcoat because of the absence of pigment and metallic or mica flake.

2.6. Curing Ovens. Electrocoat, chip-resistant primer, primer surfacer and topcoat must be cured in an oven. Paint curing is a function of time and temperature, which also varies by coating. Coating manufacturers have reduced the temperature requirements, however, further progress is highly desirable.

2.7. End-of-Pipe Controls. Bake oven exhaust volume reductions have reduced energy consumption. With low air changes, the VOC concentration is high enough to warrant incinerator destruction to be cost effective. Incineration is an excellent method of community malodor control, particularly with electrocoat bake odor.

Paint spray booth emission controls are very costly. The original capital investment can range from $10 million to $40 million. Annual operating costs for an average sized system will be $1 million to $1.5 million. Spray booth controls are considered a last resort by industry because:

> These costs are extremely high and do not add value to the process or the product.
> Equipment availability and installation problems can cause missed start-up deadlines, delaying production schedules.
> Air quality permit conditions require production stoppage when the controls are not operating properly.

At least two varieties of paint sludge drier systems are available to the coating industry. Sludge is dewatered and then conveyored through a heated chamber that dries the sludge and incinerates the volatiles driven off. The sludge is transformed into a harmless, dry, fine powder suitable as a by-product for use as a filler in sealers, asphaltic coatings, mastics, expansive concrete and some plastics.

3. The Automotive Coating System Technology of the 21st Century.

The automotive coating industry's efforts are directed towards hopes that, by the year 2000, technologies are capable of advancement to a point that its paint shops can operate with freedom from restrictions of environmental regulations. Ninety percent of the total corporate chemical releases of an automotive manufacturer are generated by assembly facilities and 90 percent of those are released by its paint shop. When armed with this knowledge, its not hard to understand why the auto industry is so attentive to environmentally beneficial technology advancements in the paint shops. Many technology gaps must be filled as we head for the 21st Century.

In Section 2, various waste streams were identified: VOC emissions exhausted from spray process, coating overspray, paint sludge containing paint overspray particles and recovered purged paint and solvent. An important waste stream that was not discussed is created by paint overspray accumulation on the booth interior and equipment surfaces. These areas must be masked or cleaned with solvents. Masking materials can become solid waste if not reusable. The cleanup solvents that cannot be recovered become either VOC emissions or are added to the paint sludge waste stream. Cleanup solvent issues are already being addressed by the U.S. EPA and industry in the short term. Therefore, this manuscript will continue to maintain a long range approach towards elimination or reduction of the paint overspray to minimize the need for maskings and cleaners.

The technology gaps that require resolution by the end of this century are all related to development of equipment required to process materials economically.

3.1. Electrodeposition Primer. Coating manufacturers have developed lead-free electrocoat materials which will soon be introduced in the industry. This will eliminate the last major environmental concern associated with this process. Continuous improvement will address other related issues such as material and energy conservation. Hopes are that a totally waterbased electrocoat material will be ready by the century's end.

3.2. Powder Primer Surfacer with Anti-Chip Enhancement. Reliable and efficient powder coating application is a primary objective of the industry. While appliance and small parts coating operations have been spraying powder finish coats for a decade, it is only recently, that the automotive industry has accomplished acceptable application of powder anti-chip and chip-enhanced primer surfacer. The chip-enhanced powder surfacer material is particularly appealing because it eliminates the anti-chip process and results in zero VOC emissions and zero air toxic emissions.

Powder coatings are substantially more costly than liquid coatings. Anti-chip and clearcoat powder film thicknesses will enhance durability and appearance at higher film thicknesses. Therefore, material costs will greatly increase. Material efficiencies of 90% must be achieved to maximize processing feasibility. This figure includes in-process recovery and reuse of oversprayed powder that is, only a 10 percent waste of material. Primer surfacer powder lends itself to reuse much more effectively than powder clearcoat. Cascading techniques of material reuse to less critical surfaces have been able to help increase the efficiency. *If powder coatings are to compete with liquid coatings, material costs must be reduced by more efficient use and reuse of the coating.*

Since the same processing issues apply, and clearcoat application is more demanding than anti-chip/surfacer, further details will be discussed in Section 3.4 for powder clearcoat.

3.3. Waterborne Basecoat. The nature of environmental improvements associated with waterborne coating technology are significant. However, waterborne technology must be examined closely to ensure that the decrease in solids and transfer efficiency can be offset by processing improvements. Although lower in solids, the material cost of waterborne coatings is equal or higher than that of high solids solventborne coatings. This loss in coating system efficiency (solids and transfer efficiency) means that to achieve equivalent coating film on the product will require an increase of 20 to 40 percent material sprayed; likewise, a corresponding increase in material cost. Material cost must be offset by other improvements, otherwise it will remain a deterrent to Pollution Prevention initiatives. To reap the environmental benefits achievable from *conversion to waterborne technology, the following technology gaps require resolution.*

3.3.1. Transfer Efficiency Improvements. Waterborne coatings are very conductive and cannot be applied with conventional electrostatic spray systems. Application efficiencies equivalent to electrostatic spray technology are critical to the long term acceptance of waterborne spray. Opportunities to achieve this objective are offered by:

> *Direct Charge Electrostatic Spray.* Previous attempts to productionize voltage block or voltage isolating systems have not been successful to date (miscellaneous voltage block designs and peristaltic pump devices). Voltage leakage drains and eventually shorts out the electrostatic charge creating a myriad of problems to maintaining a reliable charging system. To date, these leakage episodes have been difficult to solve or predict. Preventative maintenance programs have not satisfactorily resolved the system failures. Development of an early detection meter or device would be beneficial. The reliability of any charging system must surpass the reliability of the process where it is to be used. Other drawbacks have been that isolation of the charged system has created designs that increase the amount of coating lost in purging and the time required to get a good, clean purge. When coating changes are not involved, the voltage isolation system will be more successful.

Indirect Charge Electrostatic Spray. Spatial charging is currently the only production capable electrostatic charging system in full-time use in the automotive industry today. Ionization of the air around the spray pattern indirectly transfers the charge to atomized paint particles. Although this method of electrostatic charging is not as effective as direct charging, it is the only proven technology available to industry. Spatial charging has been limited to high voltage, high speed turbobell applications. Low voltage spatial charging would be desirable for robotic and manual spray applications. Safety and fire protection considerations associated with electrostatic spray systems must be satisfied in all cases.

High Volume Low Pressure (HVLP). This application technology needs further improvement before it will gain widespread acceptance. There remains a reluctance to accept HVLP spray guns due to perceived concerns of inferior finish appearance and lower rates of fluid delivery. In the short term, HVLP probably offers the greatest opportunity to improve material application efficiency. Attractiveness of this technology will also increase as plastic body panels become more popular and electrostatic benefits are negated unless costly conductive primer processes are included.

Spray Pattern Control (e.g. gentle spray, shaping air, spray gun traverse speed, etc.) Development of applicators with a soft, gentle spray and an equally distributed paint particle fan pattern could feasibly improve the material application efficiency to an acceptable level. This technology would be most desirable if designed for automatic reciprocators and robotic application equipment.

Waterborne application technology must be developed that has reliability, low maintenance costs, high application efficiency and no contribution to a waste stream.

3.3.2. Opportunities for Improvements with Materials from Paint Overspray Wastestreams.
Resins are valuable components of coatings. The paint sludge contains a mixture of resins, along with pigments, solvents, treatment chemicals, huge volumes of water and a glut of other substances. Development of a process to separate resins with an acceptable amount of impurity would allow cascading reuse of the resins in other non-critical coatings.

Solvents used in waterborne coatings must be water soluble. They are usually of higher molecular weight and slower to evaporate. Therefore, significantly less solvent will flash to a vapor state in the spray booth, meaning that more solvent:

> in the applied coating will be carried on the product to the next drying/curing stage, or
> solvent in the overspray will persist in the paint solids and sludge water. Control of the booth air exhaust stream is less effective with solventborne coatings than waterborne coatings because of these properties.

Conversely, control of the solvent in the sludge water becomes more efficient as the VOC concentrations increase. Development of a process to remove solvent from the sludge water system will compete very favorably with control equipment on the booth air stream. A continuous process that treats ten to twenty gallons of water an hour could meet the economical objectives of the coating industry.

3.4. Powder Clearcoat.
Powder clearcoat materials do not release emissions into the environment. The process operates in a closed system with minimal waste. Most importantly to industry and its customer, it will provide a more durable and more attractive paint finish.

Depth of image, distinctness of reflective image, resistance to environmental exposure/abuse and uniformity of film thickness can be improved by powder technology.

3.4.1. Optimal Application Efficiency. Application equipment for powder materials must be developed that will be capable of reaching optimal efficiencies. To date, the most promising applicator is a high voltage, high speed turbobell. Industry's goal for powder clearcoat is to reach a first-time coating efficiency of greater than 95 percent and to be able to ignore recovery and OEM reuse of the powder material. Reused powder introduces powder clusters and impurities, reducing the quality of the clear material.

Another benefit of improved first-time coating efficiency is avoidance of overspray on non-targeted surfaces, such as the product carrier and tooling. This overspray can accumulate quickly and requires frequent cleaning and maintenance. Non-intrusive methods of shielding these areas could also achieve the same objective.

3.4.2. Powder Delivery Systems. *Improved powder delivery systems are required that will deliver precise amounts and rates of material flow* from material hoppers and collection tanks to a receiver and finally to a spray applicator. Although powder coatings share many of the flow properties of liquids, many issues remain to be resolved that are caused by their differences:

> Development of instrumentation and feedback to controls is needed to measure;
>
> - Amount of material contained in receivers and rates of material flow into and out of the receivers. Correctional adjustments must be made on a continuous basis.
> - Material delivery rate through the fluid lines and to the applicator with feedback and very responsive adjustment control. This is needed to vary fluid flow rates in order to control the film accumulation on the target. It would also reduce the impact of material flow surging problems.
>
> Improved designs of the material delivery systems are needed to eliminate flow surging problems. Current systems operate on either vacuum or screw auger delivery principles.
> Development of separation techniques of reusable solids from contaminants and powder too small for impaction on the job is another opportunity to reduce waste. Recycling options for recovered, unusable solids would also be beneficial.
> *Spraybooth conversion and retrofit considerations.* Design of new systems is currently being addressed with the booth supplier community by the USCAR/LEPC. However, retrofit of existing facilities provides far greater challenges to cost effectively design a system that meets equivalent requirements. Every challenge presents an opportunity - in this case, to convert from a wet booth to a dry booth and to save investment dollars. This opportunity would encourage introduction of powder clearcoat technology on an accelerated schedule.

4. Conclusion.

The challenges of pollution prevention are clear. The needs of industry and the environment must be aligned. They must be consistent in their direction and goals. Resources for technological discovery must be combined and coordinated. We must eliminate waste with innovation that generates gainful results. Ward's Auto World reports that "by at least one knowledgeable estimate, the cost to clean up hazardous waste for automakers and their suppliers could reach a mind-boggling $500 billion over time."

Continued reliance on the command-and-control concept to achieve results will only continue creation of more waste waste of American resources that can be more wisely diverted to prosperity of the environment, the economy, industry and the consumer. In her June 15, 1993 memorandum, Ms. Browner refers to EPA program objectives for Pollution Prevention:

> *Private Partnerships*: "We will identify and pioneer new cooperative efforts that emphasize multi-media prevention strategies, reinforce the mutual goals of economic and environmental well-being, and represent new models for government/private sector interaction."
> *Technological Innovation*: "We will try to meet high priority needs for new pollution prevention technologies that increase competitiveness and enhance environmental stewardship, through partnerships with other federal agencies, universities, states and the private sector."

Protection and recovery of the environment cannot be achieved in a vacuum. Pollution Prevention is not the easy route to follow, but it is the only choice that offers everybody an improved future. It is the only option that benefits all of the players - industry, government, special interest groups, academia, research institutions and the general public. In a style reminiscent of a Vince Lombardi speech, all facets of American society must pull together with teamwork to achieve this goal.

Evidence of this emerging commitment was demonstrated at the American Electroplaters and Surface Finishing Society AESF/EPA Week, held January 24-28, 1994. Penelope Hansen, from the Office of Research and Development in Washington, D.C., stated that businesses in the United States spent $133 billion on pollution control and waste management in 1993. In the next several years, that cost is expected to surpass $300 billion. Ms. Hansen also identified that the Environmental Technology Initiative (ETI) will spend $1.8 billion over the next nine years to sponsor business attempts to apply pollution prevention technologies.

We must remove ourselves from the pollution "control" mentality of environmental protection, and step boldly and aggressively into the future with a devotion of our combined resources to accomplish permanent pollution "elimination."

5. References.

[1] American Automobile Manufacturers Association, Regulatory Impact on Paint VOC Emissions and Related Costs, Detroit, Michigan, September 17, 1993.

[2] M. A. Bindbeutel, Chrysler Corporation, Remarks to the Netherlands/Region III EPA Delegation, Newark, Delaware, June 2, 1993.

[3] G. W. Crum, Electrostatic Spraying the Three Zone Model, Powder Business Group, Nordson Corp., Amherst, Ohio.

[4] G. W. Crum, Primer on Corona-Charging Powder Guns, Powder Business Group, Nordson Corp., Amherst, Ohio.

[5] G. W. Crum, Primer on Powder Pumps, Powder Business Group, Nordson Corp., Amherst, Ohio.

[6] G. W. Crum, Understanding Load Lines, Powder Business Group, Nordson Corp., Amherst, Ohio.

[7] R. M. Dressler, E. O. McLaughlin, and R. G. Mowers, Goals and Furure Impact on the Automotive Industry, Low Emissions Paint Consortium of USCAR, presentation to SAE on March 3, 1994.

[8] B. Graves, Research and Development Forecast, Products Finishing, March, 1994.

[9] P. R. Gribble, *et al*, User's Guide to Powder Coating, The Association for Finishing Processes of the Society of Manufacturing Engineers, Dearborn, Michigan, 1985.

[10] Indiana Department of Environmental Management, Pollution Prevention for Industrial Coating, Prepared by PRC Environmental Management, December, 1993.

[11] R. Jamrog, Automotive Water-Borne Coatings, Products Finishing, October, 1993.

[12] J. Lowell, *et al*, Hazardous Waste: The Auto Industry's $500 Billion Mess, Ward's Auto World, July, 1993.

[13] National Paint & Coatings Association, The Clean Air Act and the Paint & Coatings Industry, 1992.

[14] North Carolina Department of Environment, Health and Natural Resources, Pollution Prevention TIPS, June, 1993.

[15] PPG Industries, Automotive Electrocoat Reference Manual, 1989.

[16] C. M. Smith and W. E. Brown, Elimination of VOC Emissions from Surface Coating Operations, Air & Waste, July, 1993.

[17] U.S. EPA, Pollution Prevention Policy Statement: New Directions for Environmental Protection, Administrator Carol M. Browner Memorandum, June 15, 1993.

[18] C. Weng, Electrostatic Application of Waterborne Coatings, Thesis at GMI Engineering and Management Institute, June 12, 1993.

[19] F. Wilhelm, Recycling of Water-Based Paint, Metal Finishing, October, 1993.

About the Author

Marvin M. Floer is part of Chrysler Corporation's Paint and Anticorrosion Systems Group which is responsible for all of Chrysler's paintshops. He supervises the Paint Systems Regulatory Requirements Group which has responsibility for air quality programs and all U.S. assembly plant paintshops. He holds a B.S. in Manufacturing/Industrial Engineering.

Pollution Prevention in the Metal Degreasing Industry
Stephen Evanoff, P.E., DEE
Lockheed Fort Worth Company
Fort Worth, Texas 76101
(817) 777-3772

The Montreal Protocol, Clean Air Act Amendments of 1990, and the increased commitment by both industry and regulatory agencies to pollution prevention have created directives and incentives to change the cleaning and degreasing methods used in metal working industries. The industrial sectors included in metal working are transportation, appliances, metal/alloy stock manufacturing, and polymeric composite materials. This section examines several emerging technologies and practices that offer new options for pollution prevention in this industry. For example, systems review of a manufacturing facility can identify unnecessary or redundant cleaning activities. Empty-tube bending, polymeric sheet forming films, supercritical fluids, thermal-vacuum deoiling, volatile-forming fluids, sorbent materials, and mechanical-impingement technologies, for instance, can eliminate the use of traditional solvents and, in some cases, can eliminate the requirement for cleaning altogether. In addition, innovations in ultrasonics, enhanced-depth filtration, and new cross-flow filtration technologies can improve the attractiveness of aqueous cleaning technologies. In addition to these issues, the needs and gaps in surface cleanliness measurement and aqueous cleaner analysis are identified in this section. Finally, the administrative and regulatory barriers to these changes are also outlined.

Key Words

Aqueous cleaners; composites; empty-tube bending; emulsion separation; microfiltration; polymeric-forming films; recycling; supercritical fluids; surfactant analysis; thermal-vacuum deoiling; ultrafiltration; ultrasonics.

1. Introduction

Degreasing is an integral procedure in fabricating and finishing of metal parts, during component assembly and product maintenance. Table 1 summarizes historical industrial degreasing and cleaning solvent usage. As shown, initially, flammable petroleum hydrocarbons (e.g., naphtha, kerosene, and mineral spirits) were used as ambient-temperature immersion cleaners. Beginning in the 1930s, the invention of the vapor degreaser and the bulk synthesis of chlorinated solvents (e.g., trichloroethene [TCE], tetrachloroethene [PERC], and dichloromethane [METH]) improved the efficiency, safety, and quality of metal cleaning. Thus, the chlorinated solvent vapor degreaser became the standard method for cleaning metals at manufacturing facilities. During the 1970s, concern about volatile organic compound (VOC) emissions in major urban areas led to the substitution of 1,1,1-trichloroethane (TCA) for TCE, PERC, and METH. As a result, TCA became the primary vapor degreasing solvent. CFC-113 was also implemented as a vapor degreaser at that time, but to a lesser extent.

By the late 1980s, TCA and CFC-113 were identified as stratospheric ozone depleting compounds (ODCs) and thus came under regulation. The Montreal Protocol on Substances that Deplete the Ozone Layer (1987) and subsequent amendments in 1990 and 1992 provided an international framework that bans the production of ODCs in industrialized nations effective January 1, 1996. Further, the U.S. Clean Air Act Amendments of 1990 regulate, and will ultimately restrict, the use and emissions of solvent degreasers which incorporate VOCs and "hazardous air pollutants" (HAPs). Thus, all of the historical industrial degreasing solvents are either being banned or strictly regulated. Table 2 summarizes the environmental, health, and safety issues related to historical degreasing solvents.

The emerging regulatory framework and the emphasis on pollution prevention by industry and government has led to a reevaluation of the materials and processes used to clean metals and composite materials. Industry now has a clear incentive to eliminate use of hazardous chemicals and to minimize waste and emissions where technically feasible and economically reasonable.

Table 1. Historical Industrial Solvent Use[a]

Solvent	Ambient Immersion	Heated[b] Immersion	Vapor Degreaser	Spray	Manual
Trichloroethene (TCE)	X	X	X		X
1,1,1-trichloroethane (TCA)	X		X	X	X
Tetrachloroethene (PERC)	X	X	X	X	X
Dichloromethane (METH)	X	X	X	X	X
1,1,2-trichlor-1,2,2-trifluoroethane (CFC-113)		X	X	X	X
Acetone	X				X
2-propanone (MEK)	X				X
4-methyl, 2-butanone (MIBK)	X				X
Toluene	X				X
Xylenes	X				X
Low Flashpoint Aromatic and Aliphatic Hydrocarbons and Blends[c]	X				X

[a] Derived from "1991 U.N.E.P. Solvents, Coatings, and Adhesives Technical Options Report," U.N.E.P., December 1991, p. 203.
[b] Typically in conjunction with vapor degreasing (i.e., immersion in the boiling liquid sump).
[c] Includes PD 680 Type I, PD 680 Type II, and Stoddard Solvent.

Table 2. Historical Industrial Solvent Environmental, Health, and Safety Issues

Solvent	Stratospheric Ozone Depletion	(ODP)[a]	Volatile Organic Compound[b]	Hazardous Air Pollutant[c]	Historical Soil & Groundwater Contamination	Flammable[d]	Carcinogenicity
TCE [e]	No	0	Yes	Yes	Yes	No	See Note
TCA	Yes	0.2	No	Yes	Yes	No	No
PERC [f]	No	0	No	Yes	Yes	No	See Note
METH [g]	No	0	No	Yes	No	No	See Note
CFC-113	Yes	0.8	No	No	No	No	No
Acetone	No	0	Yes	Yes	No	Yes	No
MEK	No	0	Yes	Yes	Yes	Yes	No
MIBK	No	0	Yes	Yes	No	Yes	No
Toluene	No	0	Yes	Yes	Yes	Yes	No
Xylene(s)	No	0	Yes	Yes	Yes	Yes	No
Isopropanol	No	0	Yes	No	No	Yes	No

[a] Ozone Depletion Potential (CFC-11=1.0) as listed in References 1 and 2.
[b] Per Title I of the Clean Air Act Amendments of 1990.
[c] Per Title III of the Clean Air Act Amendments of 1990.
[d] Defined as possessing a flashpoint less than 38 degrees Centigrade.
[e] The U.S. EPA has not formally classified TCE in Category B2 as a "probable human carcinogen," while the International Agency for Research on Cancer (IARC) has classified this solvent in Group 3, a substance not classifiable as to its carcinogenicity in humans.

[f]The U.S. EPA has not formally classified PERC in Category B2 as a "probable human carcinogen," while the IARC has classified this solvent in Group 3, a substance not classifiable as to its carcinogenicity in humans.
[g]The U.S. EPA has not formally classified METH in Category B2 as a "probable human carcinogen," while the IARC has classified this solvent in Group 3, a substance not classifiable as to its carcinogenicity in humans.

1.1. Scope. The industrial sectors considered in discussions in this chapter include: (1) transportation (primarily aerospace and automotive), (2) appliances, and (3) metal/alloy stock manufacturing. In addition, polymeric composites are included. Composites have been used as structural materials in conjunction with metal alloys in the transportation industry, and particularly in aerospace, since the 1980s.

This chapter discusses several emerging technologies that can potentially reduce the use of chemicals in the cleaning of metal and composite surfaces during manufacturing. These technologies are categorized according to the principles by which they operate including: (1) changes in the manufacturing system that may eliminate the need for cleaning; (2) changes in the cleaning material and process to reduce hazards, wastes, and emissions; and (3) modification in process control and waste treatment to maximize recycling and reuse of spent materials and wastes.

2. State-of-the-Art in Metal Degreasing.

The state-of-the-art and technology trends in metal degreasing were discussed in detail [1,2] previously. The general trend in industry has been to substitute aqueous and semi-aqueous cleaners for solvent vapor degreasers in simple fabrication and finishing operations [3]. Both immersion and spray methods are employed. Low vapor pressure hydrocarbons, glycol ethers, esters, ketones, and blends of these solvents are being used as substitutes for ODCs and HAPs in the following applications: (1) removal of high-viscosity metalworking fluids and straight oils; (2) removal of preservative compounds; (3) cleaning of substrate metals that are sensitive to oxidation and corrosion; (4) cleaning during equipment maintenance and service; and (5) wipe cleaning during assembly. To enhance the efficacy of aqueous and semi-aqueous cleaners, various methods of mechanical agitation, the use of ultrasonics, and the substitution of emulsifiable metalworking fluids for straight oils have been investigated and implemented, in some cases [1,2,3,6,7,8]. Methods of recycling aqueous and semi-aqueous cleaners have been tested and are also being implemented [3,7]. Industry is now examining the entire manufacturing system in an attempt to identify methods to reduce or eliminate the need for cleaning and degreasing, while concurrently attempting to optimize the substitute materials and processes.

3. Technical Opportunities.

This chapter addresses several technologies that exhibit the potential to become commercially successful within the next three to five years and identifies the technical gaps and barriers that must be overcome for implementation to occur.

3.1. Cleaning Process Elimination. Innovative methods are emerging to eliminate the use of metalworking fluids or modify the manufacturing process, thereby eliminating the need for cleaning and degreasing. Several of these methods are discussed in the following sections.

3.1.1. Manufacturing System Engineering. A comprehensive review of the entire manufacturing system at a facility can identify redundant or unnecessary degreasing and cleaning operations. Opportunities for such systemic improvements exist particularly at small- and medium-sized facilities that serve many customers or produce multiple product lines. A systematic, engineering facility review can identify these seemingly trivial opportunities. Three activities are useful in this assessment. First, quantitative block flowcharts should be developed for the overall manufacturing system and for each individual process line. Second, the levels of cleanliness required at each stage in the manufacturing process should be assessed. While the typical question that arises when replacing degreasers is, "How clean is clean?," perhaps the question ought to be "How clean is necessary?" Third, a review of material handling, transportation, and storage practices may reveal opportunities to avoid contamination. In most facilities, the information gathered through these three activities proves useful in reducing the number of degreasers and frequency of degreasing of a particular work piece. The result is a reduction in hazardous material usage with no capital investment.

Examples of changes that can be made to reduce pollution include: (1) initial routing of all heavily soiled work pieces through a single degreaser, thereby concentrating the majority of the soils in this unit and enhancing the viability of an aqueous cleaner in downstream cleaning operations as well as extending the downstream process fluid life; (2) replacing a vapor degreaser and mild alkaline cleaner process sequence with a single aqueous or semi-aqueous cleaner; (3) eliminating repeated, interim cleaning steps that are not essential to the manufacturing process cleanliness requirements; (4) placing work pieces that are not scheduled for immediate additional processing into a controlled storage area, thereby avoiding the deposition of airborne solid and liquid aerosols onto the work piece; and (5) using covered or enclosed containers for work pieces that reside in or pass through areas of high airborne contamination. Again, while these ideas may appear obvious, they represent a low-cost opportunity to eliminate cleaning. Often, an outside consultant is in the best position to provide an unbiased, unencumbered review and action plan for implementing such opportunities.

3.1.2. Polymeric Forming Films. The use of thin polymeric films applied to the die prior to performing simple bending and forming of large aluminum sheet stock has been tested in the aerospace industry and has demonstrated the potential to eliminate the use of forming fluids and the subsequent degreasing operation. Typical technical barriers that have been encountered include: (1) applications are limited to simple forming and bending operations; (2) it is difficult to place the material evenly on a surface; and (3) the adverse effects of surface contamination (e.g., moisture and solid particulate) on the work piece and die. Under the current regulatory framework, the possibility to eliminate both forming fluids and cleaning solvents and the associated compliance risks and operating costs should provide adequate incentive for industry to investigate this method. Another possible method is the use of a spray-on, peelable coating on the tool or die surface.

3.1.3. Empty-Tube Bending. This developmental technology is being investigated in the aerospace industry for the bending of aircraft hydraulic tubing. The term "empty-tube bending" refers to the absence of both an internal mandrel and forming fluids in the tooling used (Fig. 1). Currently, tube bending requires the use of forming fluids on the internal and external tube surfaces. These contaminants then must be removed in a degreasing operation [6,8]. This occurs several times during the fabrication of a tube. Companies that have tested and are interested in pursuing this technology include Boeing Company, Northrop Corporation, McDonnell-Douglas Corporation, and Lockheed Corporation. A recent Northrop Corporation assessment [9] of empty-tube bending concluded that: (1) "In terms of the aerospace industry, the empty-tube bending process is in the developmental stage"; (2) "Today, within the industry, no conclusive data exist to support or reject this process"; (3) "Given the discussions with other companies involved in developing this process, the move towards implementing this process is very favorable." Northrop Corporation is beginning

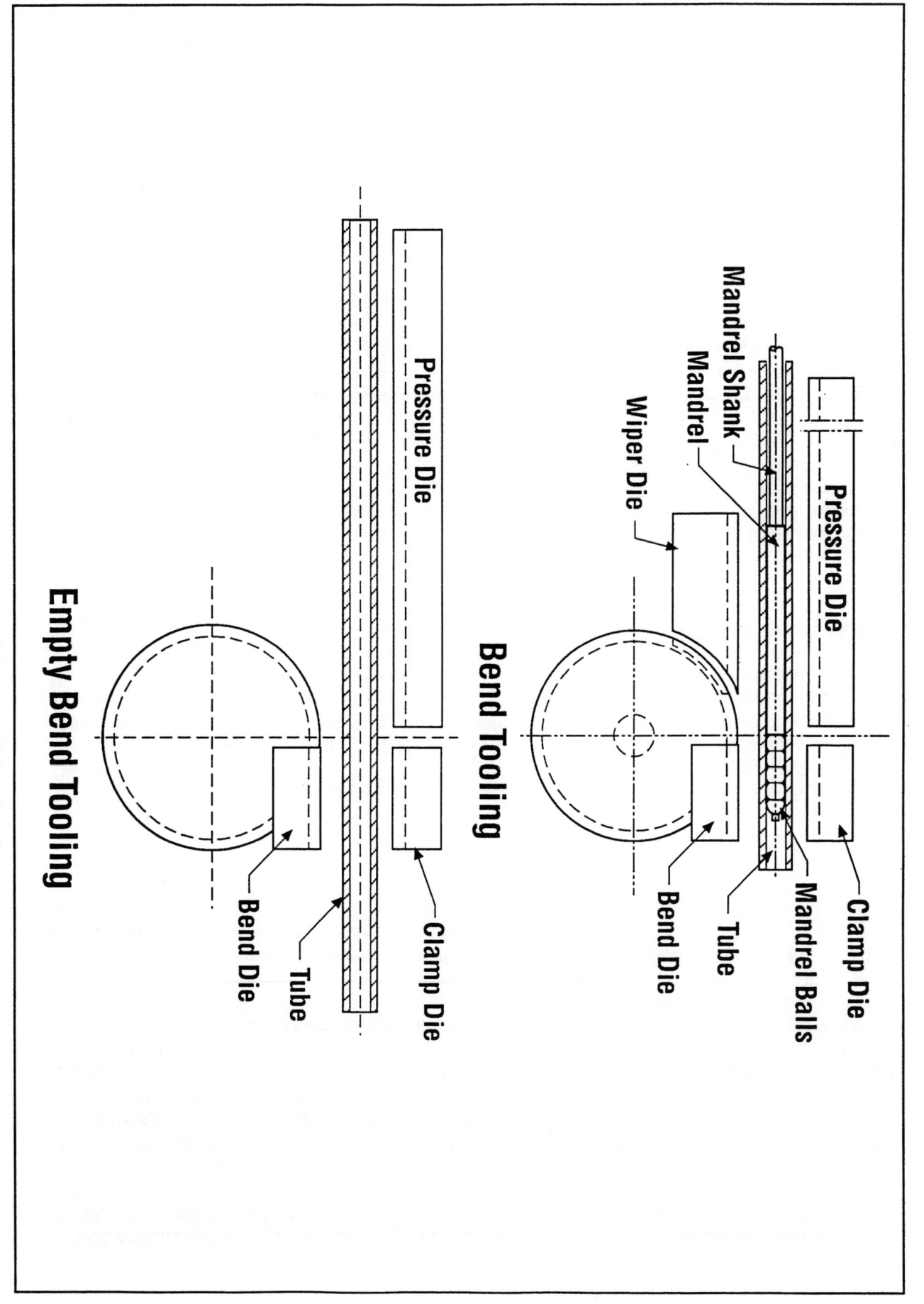

Figure 1. Empty Bend and Traditional Bend Tooling Comparison

qualification of empty-tube bending for titanium and aluminum tubing in the 1/4" through 5/8" outside diameter range [9].

Problems encountered in the work to date on empty-tube bending center on maintaining proper oval configuration of the tube in any given bend (Fig. 2). Currently, most tubing is comprised of thin-walled titanium where the "bend suitability factor" is less than 8. The "bend suitability factor" is calculated from the formula: (outside tube diameter/tube wall thickness)/"degree of bend." The "degree of bend" is defined as: bend radial length/tube diameter. Opportunities exist for proving and implementing the technology for this limited case, for expanding the technology into aluminum and stainless steel tubing, and for tubing configurations of "bend suitability factor" greater than 8. The technology could also be applied in rocket, spacecraft, missile, ship, and automobile sectors as well.

Industry has clear incentives to pursue this technology. All facilities that manufacture tubing are experiencing difficulty in eliminating the use of solvents in degreasing during tube fabrication. They consider this issue urgent due to the Montreal Protocol and Clean Air Act. Companies that have experienced some success have done so at significant capital cost, increased floor space utilization, increased manual labor, and degreaser operator labor and technical sophistication. The potential benefits of conversion to empty-tube bending include the elimination of solvents and forming fluids, reduction in labor and handling, reduced tooling costs due to simpler equipment, and reduction of internal tube surface scratches.

3.2. Material and Process Substitution. This section addresses specific cleaning materials and processes. Technologies that can potentially improve the environmental, industrial hygiene, and safety characteristics associated with cleaning are also discussed.

3.2.1. Sorbent Materials. The use of sorbent materials to hand wipe surfaces of simple geometry that are contaminated with low-viscosity liquid soils can eliminate solvent cleaning during fabrication and assembly. Sorbent material candidates include: (1) coarsely woven polypropylene pads, (2) peat fiber, (3) starches derived from various plant matter, (4) diatomaceous earth, (5) kaolin, (6) talc, and (7) cellulose fiber. All items listed, except for the propylene pads and cellulose fiber, can be applied either directly upon the work surface as a bulk material and removed using a cloth wipe or a pad, or integrated into an open-weave wipe pad configuration and used to wipe the surface. These materials have been tested and implemented on a limited basis within the aerospace sector [5,10].

This approach could have broad applications, particularly in small- and medium-sized facilities that lack capital facility funding and in industry sectors that manufacture products with less rigorous surface cleanliness requirements. Potential technical barriers that remain to be addressed include limitations in applicability to various soil types, residual contaminant and sorbent on work pieces after cleaning, and abrasion of the work piece, in the case of inorganic sorbents. Other issues to be addressed include work technique, user acceptance, process time, material costs, and waste disposal quantities and costs.

3.2.2. Mechanical Impingement [1]. The use of highly pressurized inert gases, nitrogen, carbon dioxide, and air for displacement of particulates from surfaces has demonstrated application in precision cleaning. Media blasting using starch granules, solid carbon dioxide pellets, and sodium bicarbonate granules has been tested and has proven effective in selected metal and precision cleaning applications [1]. Key technical barriers to be addressed include (1) size and sophistication of equipment; (2) capital cost of equipment; (3) applicability to large workpieces and those with complex shapes; (4) surface cleanliness in cases where soils include polar organics, non-polar organics, and colloidal solids; and (5) safety and industrial hygiene. The biggest potential benefit is the elimination of solvents and reduction and possible elimination of waste and regulated emissions. These techniques will be of greatest interest to organizations that manufacture smaller, precision metallic components and instrumentation.

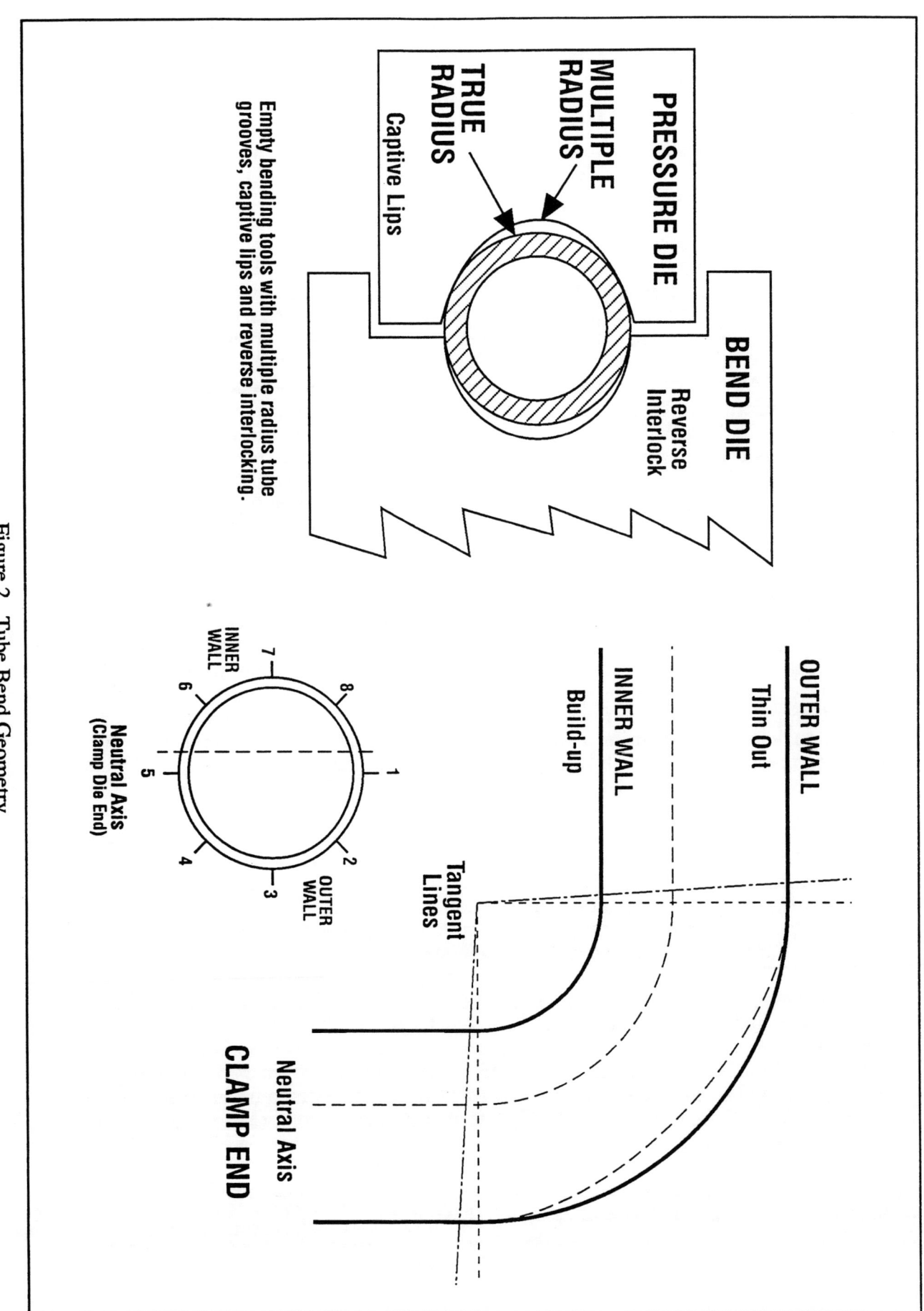

Figure 2. Tube Bend Geometry

3.2.3. Thermal Vacuum Deoiling [1,2]. Thermal vacuum deoiling is a concept that incorporates hydrocarbon-based forming fluids capable of being evaporated from workpieces in a heated vacuum chamber. The vapor-laden air stream then is passed through a condenser where the forming fluid is recovered and removed from the system. This technology will have limited applications because it can only be used in situations where the soil type and loading are consistent and the part size and geometry are simple.

3.2.4. Supercritical Fluids. Supercritical fluids (SCFs) exist in conditions above their given critical values for temperature (Tc) and pressure (Pc). The critical region is generalized as:

$0.9 < Tr < 1.2$ and
$1.0 < Pr < 3.0$

where: Tr and Pr are the reduced temperature and pressure, respectively, and

$Tr = T/Tc$
$Pr = P/Pc$

Typically, gases are compressed into the supercritical region to create an SCF. In this state, many fluids exhibit a significant increase in dissolution of organic contaminants. Also, SCFs have high diffusivities and low viscosity and low density relative to chlorinated solvents.

In practice, a work piece is placed into a closed chamber. The gas that is to perform the soil dissolution is heated and compressed into the supercritical region, pumped into the cleaning chamber, and then decompressed in a separate vessel where the soil and gas are separated. A block flowchart of a simplified SCF system is shown in Fig. 3 [11]. Supercritical carbon dioxide [1, 11] and supercritical ethane [12] have demonstrated efficacy in industrial cleaning applications involving small, critical instrumentation and components. This technology shows great promise in all cases where capital cost is acceptable or where no other technically feasible method is available, as in the case of certain precision instrumentation.

3.2.5. Aqueous Cleaners. Aqueous cleaners are increasingly replacing solvent vapor and ambient temperature immersion degreasing in metal and composite material manufacturing. Aqueous cleaners can remove hydrocarbons, polar organic, and inorganic soils, and colloidal solids, when proper operating conditions exist. These conditions typically include temperatures in the 50-70 degrees Celsius range, concentrations well in excess of the critical micelle concentration, and high levels of mechanical agitation or some other form of cleaning enhancement in the process. Without such conditions, the cleaners would not perform acceptably in the removal of the aforementioned soils. In fact, even at extreme conditions of temperature, concentration, and agitation, high-viscosity soils and semi-solid soils are rarely removed without causing etching or some surface corrosion to the work piece.

Innovations in surfactant blending and cleaner formulation in aqueous and semi-aqueous cleaners could lead to more aggressive products that could be employed at lower temperatures. Phosphates and silicates are detergent builders and, as such, are essential components in aqueous cleaner chemistry. However, cleaners formulated without silicates would reduce concerns about surface residues, which might create a surface condition averse to coating or bonding. Further, the minimization of phosphates in cleaners is desirable, as phosphate discharges may become regulated in some regions in the future [13]. The literature [14] describes other detergent builders that are available such as sodium citrate, sodium hydroxide, sodium carbonate, sodium bicarbonate, and sodium tetraborate. None of these materials perform as well as phosphate and silicate as detergent builders and sequestering agents.

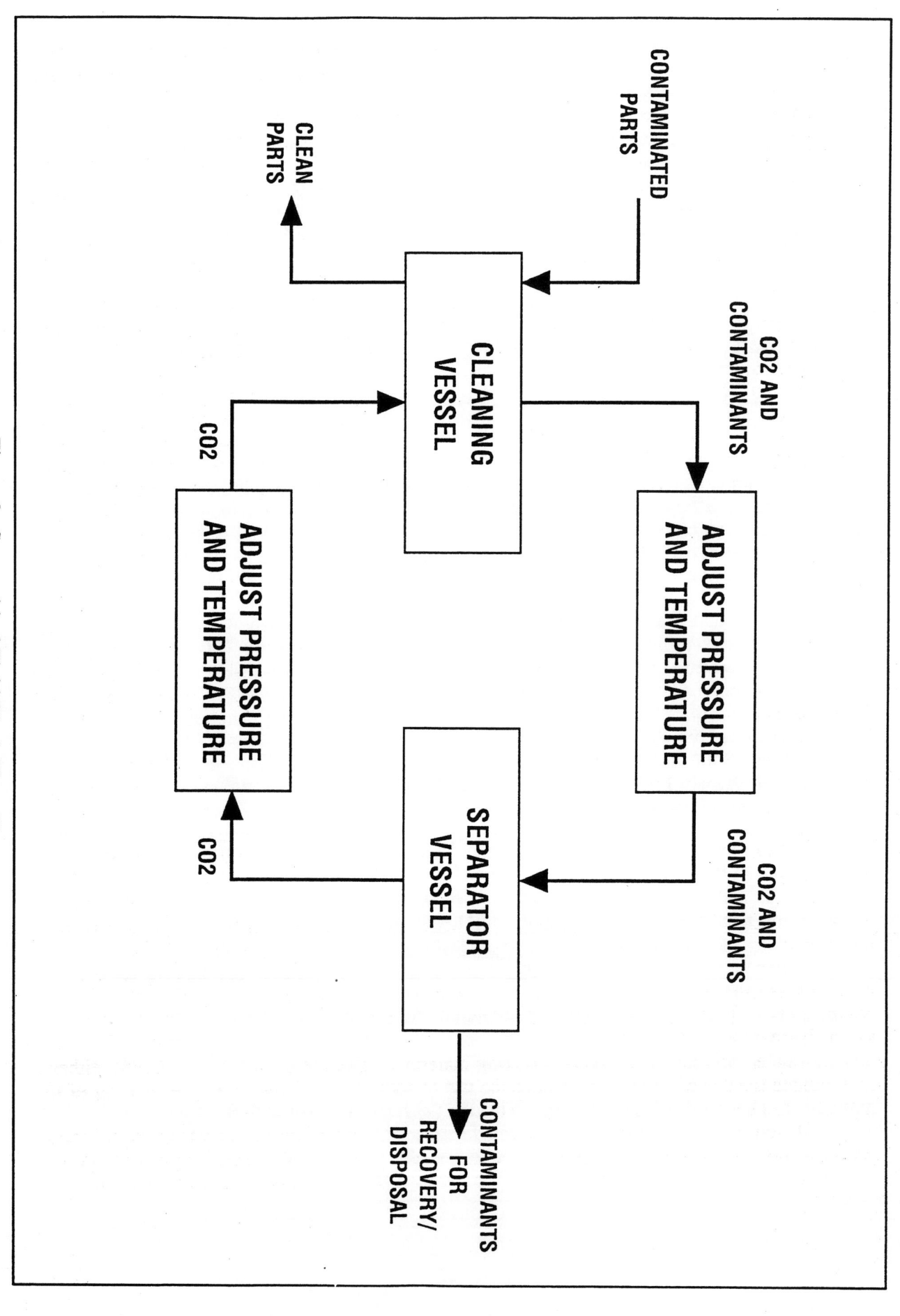

Figure 3. Supercritical Fluid Block Flow Chart

In summary, more aggressive cleaners without increased corrosion risk, cleaners that operate at lower temperatures, and cleaners with reduced phosphate and silicate content would be of interest to regulated industries.

A major barrier to implementation of such cleaners is that qualification and certification of the products will be expensive and time-consuming. Also, many companies will have already substituted an aqueous cleaner within the past several years and, therefore, may hesitate to revisit this issue so soon after making a similar change. The most successful vendors of cleaner technology have been those that allow their product research chemists to work directly and freely with the customers' chemists and who spend time at the customers' facilities learning the nuances of their operations.

3.2.6. Low Vapor Pressure Solvents. In cases where high-viscosity, semi-solid, and solidified soils are present, an organic solvent or blend may be required. Low vapor pressure blends of hydrocarbons, esters, glycol ethers, and ketones are emerging and serving in this capacity as wipe and ambient temperature immersion cleaners. Under current U.S. environmental regulations, these products must be free of ODCs and HAPs, possess a vapor pressure less than 45 mm Hg, be non-flammable, and pass rigorous toxicological standards in order to be attractive to a manufacturing facility. Typically, these solvents emit objectionable odors and pose potential hazards for the swelling or degradation of polymeric and elastomeric materials. Some enhanced drying may also be necessary. Products that meet the technical requirements, emit low odor, and reduce risks would be attractive to manufacturers. The greatest barriers to implementation of such products are the costly, time-consuming qualification testing that would be required; user resistance to change based upon the more application-specific cleaners; the increased sensitivity to technique; and sensitivity to solvent odors.

3.2.7. Cosolvent Systems. Systems that incorporate alcohols or other organic solvents with hydrofluorocarbons or perfluorocarbons have been developed and implemented in some cases [1]. The solvents must be immiscible with the alcohol or solvent serving as a heated or boiling liquid cleaner and the PFC or HFC serving as the rinsing and drying medium. The HFC and PFC also suppress the flammability problems of the heated solvent.

These systems are useful in cases where corrosion- or oxidation-sensitive alloys are used and with soils that require heated organic solvents. However, PFCs are under scrutiny for strict regulation and possible ban in these applications [15]. Innovations in cosolvent formulation and cleaning system design safety and simplicity are of interest to manufacturers facing the special problems that require this technology. This is a relatively small market, but one with few current options.

3.2.8. Maskant Materials. The use of high-boiling temperature paraffinic waxes as masking materials during plating of new products and during rework has been prevalent in the transportation sector. The principal items that require masking and plating include ferrous metal alloys used in landing gear, jet engine components, and load-bearing surfaces and components. Typically, wax maskants are removed with tetrachloroethylene (PERC) in a vapor degreaser. Recent work in the aerospace industry has focused on substitution of nonchlorinated solvents or aqueous dewaxing materials. The use of peelable maskants, either to eliminate the use of solvents or reduce the use of solvents to a hand wipe after peeling of the maskant, has been described in a recent Toshiba Corporation presentation [16].

Development of a product that meets the adhesion and chemical resistance requirements for maskants, but is peelable, appears to be technically feasible and of great interest to the transportation sector. The need to reduce the use of chlorinated solvents should enhance industry incentive to consider such products. Also, labor and operating costs could potentially be reduced. Barriers include the rigorous testing and qualification regime to which candidate products are subjected.

3.2.9. Preservative Compounds. The use of preservative compounds, commonly called "Cosmolene," is widespread in the metal working industry, particularly when components and products are shipped across oceans. The automobile industry uses asphaltic preservative compounds, to coat surfaces that may corrode, before shipping. A peelable preservative compound would eliminate the need for use of aggressive solvents such as chlorinated solvents, terpene hydrocarbons, and naphthas. Alternative materials would have similar properties and applicability as those described in "Sec. 3.2.8 Maskant Materials." MIL-C-16173 lists the specification requirements for preservative compounds.

3.2.10. Forming Fluids. A Japanese manufacturer reports that volatile forming fluids that do not require solvent cleaning, thermal vacuum deoiling [1], or other methods to remove the forming fluid residue after tube bending, are being developed and implemented. Such products would offer the potential to avoid capital and operating cost increases associated with alternative cleaning technology and provide a means of eliminating use of solvents in cases where empty-tube bending is not technically feasible. These cases would still retain the environmental compliance risks associated with VOC emissions. However, VOC emissions from such an operation could be minimized and captured if necessary. The somewhat conflicting technical requirements for an organic compound that retains lubricity under high pressure and high temperature conditions, yet which is volatile, make the potential of success of such a technology questionable.

3.2.11. Ultrasonics. The use of ultrasonics to enhance the efficacy of cleaning has been evaluated and reported as companies search for methods to eliminate solvent degreasing [1,2,6,8]. A Japanese firm has reported on several occasions on the enhanced cavitation technology which relies upon multiple frequencies (e.g., 28 kHerz, 100 kHerz, and 200 kHerz) in aqueous and solvent immersion cleaning systems. A recent paper [17] describes the use of this technology for cleaning and deburring small and precision metal parts. The use of ultrasonics and vacuum drying offers the potential to simultaneously eliminate solvents and improve cleaning and deburring quality. Significant testing and capital costs would likely be involved in affecting this change. Barriers to ultrasonics techniques are: (1) applicability to large systems, (2) homogeneous distribution of energy, (3) damage or corrosion to work piece, (4) noise levels, and (5) capital and operating costs.

3.3. Waste and Emissions Reduction. This section addresses technologies that can reduce waste and emissions by improving the process efficiency or through recovery, recycling, and reuse.

3.3.1. Analytical Methods. The analytical methods used to determine levels of cleanliness in performing metal cleaning have come under scrutiny as companies test and implement substitute cleaning materials and processes. Most simple methods are qualitative, while the quantitative methods tend to be much more expensive, time and labor intensive, and technically sophisticated. Several studies and handbooks have listed, evaluated, and compared the efficacy and application of these methods [3,6,7,8,18]. In particular, the replacement of solvent degreasing of tubing that carries oxygen or other highly reactive chemicals and oxidizers has highlighted a need for a simple, quantitative cleanliness test method for organic and inorganic soils [6,7,8]. The optical scanning electron emission (OSEE) technology which uses ultraviolet light to excite surface electrons and then measures the energy emitted from the surface has shown promise as a simple, quantitative, analytical method that is relatively inexpensive and can generate large amounts of data quickly in a factory or laboratory setting. This method and other innovative applications of existing quantitative methods for measuring organic and inorganic soils, and both solid and non-volatile liquids, are of great interest to all facilities that are investigating alternative cleaning methods. Any cleanliness method should be applicable to small- and medium-sized metal working firms and, as such, should be simple, repeatable, and quantitative.

Also, analytical methods for determining the condition of aqueous cleaners (e.g., soils and surfactants) are inadequate for reasons similar to those for determining cleanliness. Measurement of pH and titrimetric methods can accurately measure the ionic detergent builder concentration, but do not measure the presence of the surfactants and wetting agents, that are critical to cleaner performance. High-performance liquid chromatography (HPLC) [6] and Fourier Transform Infrared Spectroscopy (FTIR) have been applied in some limited cases. FTIR and visible-light spectroscopy have been employed in very limited cases for soil concentration determination. These methods can be applied in a manner that provides accuracy and precision of analysis. To date, literature indicates that this has only been accomplished in controlled laboratory and pilot tests and when the soils and surfactants are limited to one.

3.3.2. Electrostatic Bag Filters. A recent paper [19] describes the application of electrostatic charges to polymeric bag type filters to enhance hydrocarbon oil separation and removal from aqueous detergent solutions. The authors claim that this method facilitates recycle and reuse of cleaners and rinsewaters. This method could also further enhance the attractiveness of aqueous cleaners and solvent replacements. Further, if effective, this method has a lower capital cost, and is a technically simple alternative to microfiltration or ultrafiltration. Interest in this method would be widespread in the metal working industry. Technical concerns with this method focus on the need to understand the impact, if any, of adding an electrical charge to an ionic solution and the possible adverse effect upon the cleaner.

3.3.3. Electrochemical Emulsion Separation. A recent Japanese paper [20] describes the potential for using a low-voltage, high-frequency electric field to oil detergent emulsions in a manner which causes flocculation and helps to separate the emulsion. The author claims that, "Under such a condition, the emulsion particles are flocculated rapidly along the line of the electric force and then separated effectively." Concerns with introducing electrical fields to aqueous solutions used in metal cleaning include the potential for enhanced corrosion of substrate and damage or degradation of the cleaning fluid constituents.

3.3.4. Microfiltration and Ultrafiltration. Cross-flow membrane technologies are now being used at several industrial facilities to recycle soiled aqueous alkaline cleaners [3]. These technologies include ceramic membranes and polymeric membrane systems. The principal technical challenges to full-scale implementation of this technology are the development of operating parameters that maintain an acceptably high permeate flux (i.e., regenerated cleaner) while achieving maximum concentration of the soil in feed (i.e., soiled cleaner), minimizing the impact of fine colloidal solids on membrane performance and, most importantly, demonstrating adequate separation of soil and recovery of cleaner constituents by a reliable analytical method.

3.3.5. Composite Material Recycling. Composite materials began to replace metals as structural components in the transportation industry in the 1980s [21]. Scrap broadgoods and cuttings from the lay-up and fabrication of graphite-epoxy and fiberglass type composite materials are not regulated under the Resources Conservation and Recovery Act (RCRA) for hazardous characteristics. Nevertheless, the increase in waste composites has generated interest within the aerospace industry for recycling composite scrap into other products. The Great Lakes Composites Consortium and the Aerospace Industries Association (AIA) are investigating the processing of uncured composite broadgoods scrap from aerospace manufacturing activities into components for aerospace ground support equipment and vehicles. Composite scrap from various aerospace sources is also being investigated. McDonnell-Douglas is leading testing to determine the range of properties and conditions for composite scrap that is acceptable for processing into alternate, non-structural articles. Figure 4 is a flowchart of the composite recycling concept under development. Simply stated, scrap pre-impregnated broadgoods would be separated from paper backing and other materials, ground into small pieces, placed into molds of various shapes, and cured in an autoclave in a manner similar to new aerospace material.

Several parameters require investigation and definition for this concept to become technically feasible and economically attractive. These parameters include: (1) the physical condition of scrap material as received from generators, (2) the range of physical and chemical properties of the scrap broadgoods as compared to the new material, (3) the shipping and handling requirements for scrap broadgoods, (4) the minimum quantities of scrap broadgoods required, (5) the degree of commingling allowed between various scrap materials that is acceptable, and (6) the specifications for batch processing scrap broadgoods.

4. **Administrative and Regulatory Barriers to Innovation.**

The administrative and regulatory barriers to innovation can be summarized in the following broad categories: (1) command-and-control regulations, (2) recordkeeping and reporting requirements, (3) conflicting regulations, and (4) specifications and standards.

Environmental regulations should establish a framework that emphasizes reductions in sources and emissions without specifying methods and approaches. Regulators should realize that generalizations do not broadly apply across industry sectors. Instead, affected parties should agree to the desired pollution prevention outcome, the technically feasible options and limits, and then proceed accordingly. In exchange, industry should commit to integrated pollution prevention strategies and provide the resources necessary to develop and implement technical innovations.

Recordkeeping and reporting requirements add to the already excessive burden on environmental industry professionals. In the past, most recordkeeping and reporting did not contribute to pollution prevention or the development of new or innovative technologies. Recordkeeping should be limited to that necessary to verify compliance with existing laws and regulations. Records should be constructed in a manner that supports development of baselines for future negotiations between industry and regulators.

State and federal regulations have the potential to create conflict for industry in compliance. For example, federal regulation of ODCs may require substitution of a VOC-containing product that is heavily regulated by a state agency. Since ODC is not regulated at the state level, restrictions may be excessive on VOC sources, leaving industry no acceptable choice. Again, partnerships between regulators and industry and a focus on pollution prevention rather than command-and-control regulation can help resolve these conflicts.

Industrial specifications and standards, particularly within the defense industry, create a tremendous resistance to change the materials and processes used to manufacture and maintain products. Typically, alternate products require detailed, rigorous testing and qualification for each individual organization, facility, and application. This results in much unnecessary and redundant testing and long spans of time between testing and implementation. In some cases, this large administrative burden will result in the retention of the existing qualified process and the addition of end-of-pipe control technology instead of the implementation of a pollution prevention process.

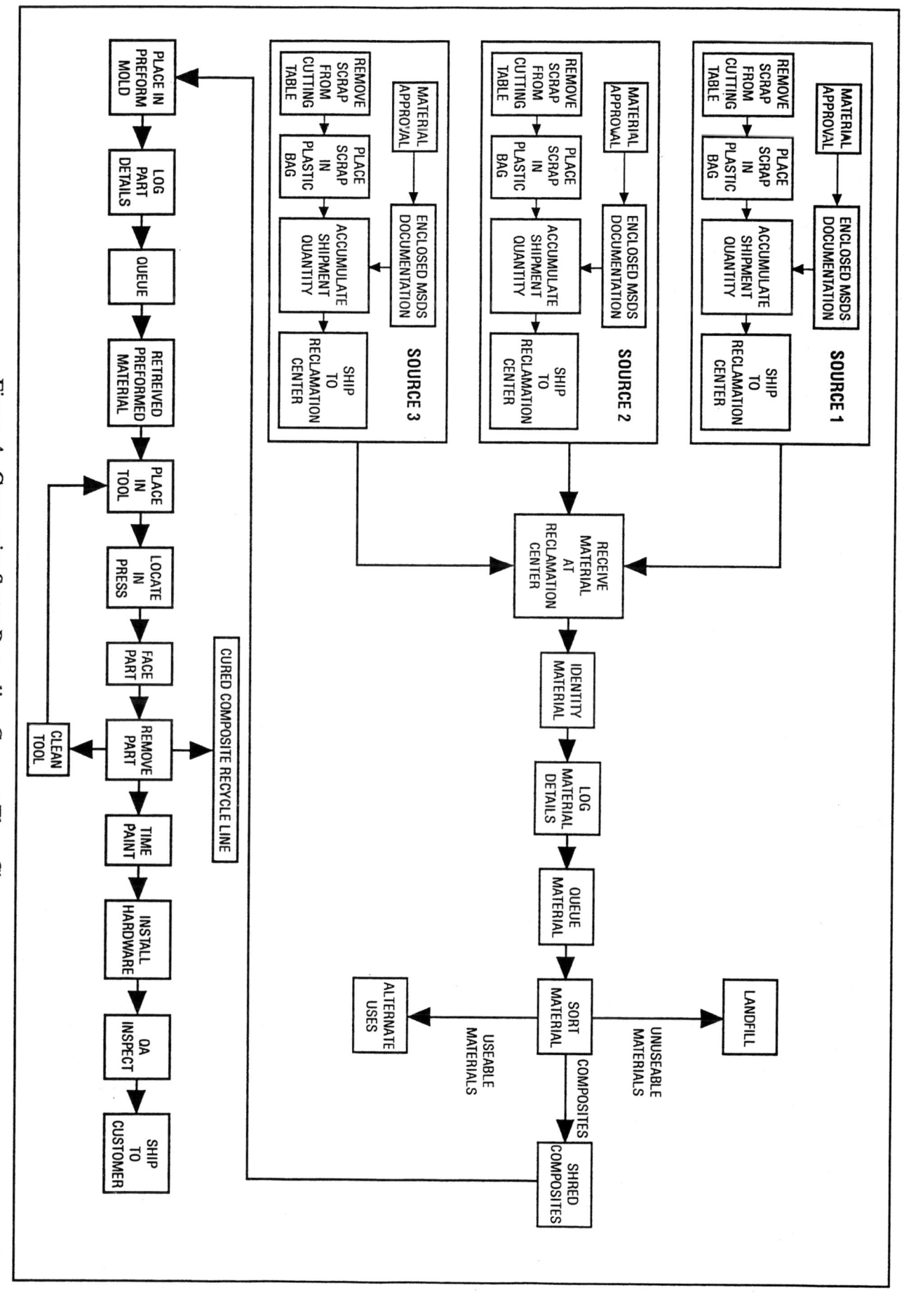

Figure 4. Composite Scrap Recycling Concept Flow Chart

5. Conclusions

The national and international regulatory frameworks being implemented today and the increased emphasis on environmental stewardship within industry are driving metal working and composites manufacturing facilities to eliminate the use of high volatility solvents in cleaning and degreasing. Most facilities are considering and implementing aqueous, alkaline cleaners and, to a lesser extent, semi-aqueous cleaners and low vapor pressure organic solvents and blends. This conversion will accelerate through the end of the century.

The substitution of "no clean" methods and technologies offers the greatest long-term promise to pollution prevention in the metal and composites industries. The most thoroughly tested and developed technologies that offer significant benefits to industrial users include empty-tube bending, polymeric films for metal forming, mechanical impingement cleaning, and supercritical fluid cleaning.

Other promising but less thoroughly tested concepts for practicing pollution prevention include thermal vacuum deoiling, volatile forming fluids, electrostatic enhancement of depth filtration, and electrochemical emulsion separation.

Multiple frequency ultrasonics and the recycling of scrap composites are being used on a limited basis. The use of these methods should increase with time and become integrated into standard industrial practice.

A broad need exists for technologies to treat and either recycle or reuse soiled aqueous cleaners. Several such technologies appear promising and are beginning to be implemented at industrial facilities.

The largest single technical barrier to be overcome is the need for a simple method of determining cleanliness of surfaces and the condition of aqueous cleaners that have been treated and are desired to be reused. In most cases, the technologies required are available and require an organization to integrate the individual materials and methods into a service or product line.

The successful firm will combine a consulting service to review the entire manufacturing system with an integrated process technology approach. The ability to provide modular, portable, rugged equipment and to combine modules in a manner that offers a superior process are essential to success in this manufacturing sector.

The regulatory requirements of the Montreal Protocol and Clean Air Act Amendments of 1990, along with industry's increased commitment to practice pollution prevention have created unprecedented opportunities for the use of new cleaning materials and processes and related equipment and analytical methods in all metal-working operations over the next three to five years.

Acknowledgment. Special thanks to Dr. Tom Woodrow of Lockheed Fort Worth Company for his recommendations and suggestions in writing this chapter

.6. References.

[1] UNEP Technical Options Committee, 1991 UNEP Solvents, Coatings, and Adhesives Technical Options Report, United Nations Environment Programme, December 1991.

[2] ICOLP Technical Committee, Alternatives for CFC-113 and Methyl Chloroform in Metal Cleaning, EPA/400/1-91/019, U.S. Environmental Protection Agency, Washington D.C., June 1991.

[3] H.J. Weltman and S.P. Evanoff, Replacement of Halogenated Solvent Degreasing with Regenerable Aqueous Cleaners, 46th Annual Purdue Industrial Waste Conference Proceedings, Lewis Publishers Inc., Chelsea, Michigan, USA, 1992, pp. 851-871.

[4] H. J. Weltman and T. L. Phillips, Environmentally Compliant Wipe-Solvent Development, SAE Technical Paper Series #921957, Society of Automotive Engineers, Inc., Warrendale, Pennsylvania, USA, 10 pages.

[5] T. L. Phillips, H. J. Weltman, S. P. Evanoff, and B. D. Campbell, Development and Implementation of CFC-Free Manual Cleaning Solvents at Air Force Plant No. 4, 93-RP-152.04, 86th Annual Meeting of the Air and Waste Management Association, Denver, Colorado, USA, June 1993.

[6] T. A. Woodrow, K. M. Koepsel, and J. O. Karnes, "Aqueous Alkaline Degreasing of Aircraft Tubing, The 1993 International CFC and Halon Alternatives Conference, Washington D.C., 21 October 1993.

[7] K. B. Evans and K. J. Schulte, Alternatives to Vapor Degreasing of Solid Rocket Metal Components, Proceedings of the 8th Annual Aerospace Hazardous Materials Management Conference, Chandler, AZ, October 1993.

[8] S. Bose and A. Wijenayake, "Environmentally Compliant Tube Cleaning, Proceedings of the 8th Annual Aerospace Hazardous Materials Management Conference, Chandler, AZ, October 1993.

[9] R. Bowen, Northrop Corporation Facsimile to S. Evanoff, 14 December 1993.

[10] E. Groshart, Boeing Corporation Correspondence to Lockheed Environmental Systems and Technologies, 1993.

[11] R. Novak, Cleaning of Precision Components with Supercritical Carbon Dioxide, Proceedings of the 1993 International CFC and Halon Conference, Washington D.C., October 1993.

[12] D. Hunt, Using Different Technologies to Solve Unique Precision Cleaning Problems, Proceedings of The International CFC Alternative Cleaning Technology Conference, Yokohama, Japan, 15-16 December 1993.

[13] P. Diessel, J. H. Stabenow, and W. Trieslet, Wirkungsweise von Copolycarboxylaten in Washmitteln, Tenside Surfactants Detergents, Carl Hanser Verlag, München, 1988.

[14] CPI Firms Aim to Clean Up with New Detergents, Chemical Engineering, 23 May 1977, pp. 104-108.

[15] 58 FR 28094, Significant New Alternatives Program, Proposed Rule, 12 May 1993.

[16] S. Matsui, Opportunities for Cooperation on Aircraft and Aerospace Projects, Presentation to United Nations Environment Programme Technical Options Committee, Yokohama, Japan, 14 December 1993.

[17] A. Takabayashi, Metal Cleaning Application, The International CFC Alternative Cleaning Conference, Yokohama, Japan, 15-16 December 1993.

[18] Wood, W. G., Surface Cleaning, Finishing, and Coating, Metals Handbook Ninth Edition, Volume 5, American Society for Metals, 1964, pp. 20-21.

[19] O. Kimihiko, Treatment Technology of Washing by Using Electrostatic Filters, The International CFC Alternative Cleaning Conference, Yokohama, Japan, 15-16 December 1993.

[20] Y. Kikuchi, New Technology to Separate the Oil from O/W-Type Emulsion System, The International CFC Alternative Cleaning Conference, Yokohama, Japan, 15-16 December 1993.

[21] S. Evanoff, Material and Process Change: An Aerospace Industry Perspective, Rethinking the Materials We Use: A New Focus for Pollution Policy, K. Geiser and F. H. Irwin, Editors, 1993, pp. 113-120.

About the Author

Stephen P. Evanoff is Manager of the Lockheed Corporate Environmental, Safety, and Health Department in Las Vegas, Nevada and is the former Manager of the Environmental Resources Management Department at Lockheed Fort Worth Company. He serves as a member of the United Nations Environment Programme Solvents, Coatings, and Adhesives Technical Options Committee. He is a Registered Professional Engineer, a Diplomate of the American Academy of Environmental Engineers, and a Registered Environmental Manager. He has co-authored 24 technical papers, articles, and book chapters on pollution prevention technologies and holds one patent.

Pollution Prevention in the Metal Finishing Industry
Kevin P. Vidmar
Division Manager, Environmental Affairs and Plant Services
Stanley Fastening Systems
East Greenwich, RI 02818
(401) 884-2500

Products from the metal finishing industry are important to our everyday lives, and this large industry generates significant amounts of pollution to the air, water and land. Water discharge regulations have helped clean up the receiving streams, while the pollution prevention thrust within more recent years has reduced the types and total amounts of pollutants getting to this discharge point. However, there still exists substantial amounts of progress to be made.

In this chapter, typical pollution prevention strategies and technologies within metal finishing are discussed, providing a present state of the art. Opportunities exist for future pollution prevention in: 1) solution chemistry advances, 2) substrate substitution, 3) end use, green product design, 4) application of existing technology (or its combination) to new or untried problems, 5) centralized processing, 6) metal hydroxide management facilities, and 7) vendor processing. Numerous regulatory barriers, typical market concerns, and past problems associated with the introduction of new technologies, provide obstacles to these opportunities. Pollution prevention grants, financial incentives for unproven technology, or other positive means to encourage trailblazing, will help alleviate barriers. An improved and more workable, yet still strict, regulatory atmosphere is also needed.

Key Words

Metal finishing; plating; pollution prevention, recycling; wastewater treatment; waste reduction.

1. Introduction

The historical roots of the metal finishing industry can be dated back to the 19th century. For example, chromium plating can be traced to 1856, while the first significant commercial plating of chromium took place in the 1920s [1]. The plated products of this industry are used in our everyday lives. From transportation to the tools of our trades, we use a wide variety of products with surfaces whose finish has been changed either mechanically or chemically to impart different characteristics or appearance. A recent survey estimated there are over 13,000 metal finishing facilities within North America that generate substantial amounts of wastes [2]. These may be potential pollutants as a result of metal finishing operations emitted into the air or water, or placed on the land.

The goal of this chapter is to briefly describe: 1) the metal finishing process, 2) general waste characteristics and generation, 3) the succession of pollution prevention that has occurred within metal finishing, ending with present state-of-the-art, 4) areas of future pollution prevention opportunities, 5) limitations to future pollution prevention, and finally 6) ways to overcome and address these barriers.

2. What is Metal Finishing

This section discusses the purpose and general types of metal finishing. For more detailed specifics on metal finishing unit operations, including the specific chemicals, hardware, and techniques involved can be found in Refs. [3] and [4].

2.1 Types of Operations. Metal finishing typically involves a substrate, such as metal or plastic, that undergoes a variety of chemical, mechanical, or electrochemical operations to provide specific surface properties to the substrate parts.

2.1.1 Mechanical Operations. These operations alter the surface by physical means and include processes such as polishing, buffing, blast finishing, and tumbling (mass) finishing. These processes do not require series of tanks with solutions and rinses and generate minimal wastes as compared to chemical or electrochemical processes.

2.1.2 Chemical Operations. These operations usually involve cleaning, surface preparation, or activation prior to electrochemical operations or post-plating operations such as conversion coating. Chemical operations include degreasing, acid or alkaline cleaning, etching, or pickling, and conversion coatings such as chromating. Examples of chemical operating solutions include sulfuric acid and hydrochloric acid.

2.1.3 Electrochemical Operations. These operations represent the actual plating and involve depositing one metal (in solution) over another (substrate) by utilizing an electric charge. Plating can also be performed without the use of current, as with electroless (autocatalytic) plating. Benefits of metal plating include [5]:

> Protect the base metal (zinc or cadmium on steel)
> Improve appearance (nickel and chromium on steel)
> Improve hardness and wear resistance (chromium or electroless nickel on steel)
> Lower contact resistance and increase reliability for electrical contacts (gold on brass or copper)
> Improve solderability and/or weldabiltiy (tin on brass, electroless nickel on steel)

Provide a better base for other finishes (nickel under gold or chromium)
Improve lubricity under pressure (silver on bronze)
Strengthen the base and render it more temperature resistant (copper, nickel, and chromium on plastics)
Act as a stop-off in heat treating (copper on steel for carburizing)

Electrochemical metal finishing operations and associated chemical operations generate the largest volume of wastes within the industry.

2.1.4 Additional Operations. Other operations are used to place one metal over a substrate of a different material. For example, hot dipping is utilized to immerse part or all of a workpiece into a molten bath of another metal. Vacuum metalizing methods such as vacuum evaporation, ion plating, and sputtering also provide metallic layers on substrates and have been used in a variety of applications to replace conventional plating. References [4] and [6] provide details on the specific requirements of this technology. Vacuum processes offer the advantages of less pollution, less chemical usage, and elimination of the need for mechanical operations such as polishing or buffing. The disadvantages of vacuum technology is its initial cost, its limited substrate coverage on various parts, and the typical required use of a base and topcoat (lacquer or other organic coating) which can cause air pollution issues.

2.2 Typical Metal Finish Plating Operation. In simple terms, a typical metal finishing operation consists of 1) surface cleaning and preparation of the parts, 2) processes to change the surface or properties to those desired (such as application of one metal over another through electrochemical or electroless plating), and 3) rinsing, and possibly mechanical finishing operations as necessary.

Figure 1 illustrates the integration between the required operations and the steps involved in a typical plating system, in this case using alkaline zinc. All pieces proceed through the plating line either on racks or in barrels. First, the parts are cleaned of dirt and oils using an alkaline cleaning step followed by one or more rinse tanks. Next, pieces are rinsed in tanks containing solutions such as sulfuric acid to counteract the alkaline properties of the soak clean, to remove surface scales and oxides, and to activate the surface for the actual plating process to follow. After one or more rinse steps, the actual zinc plating occurs. Following plating, the parts proceed through one or more rinses and then through a mild acid solution to neutralize the alkalinity of the zinc plating solutions. The parts then move though the chromate conversion coating process, which enhances the corrosion protection provided by the zinc coating. Final rinsing and drying steps follow the chromate conversion step.

3. General Waste Characteristics

Liquid wastes typically generated by the metal finishing industry include [7]:

Spent plating baths include spent electroplating or electroless baths, and baths that have been contaminated with other metal finishing materials.
Spent process baths include cleaners and etchants that no longer serve their purpose.
Strip or pickle baths include nitric, sulfuric, hydrochloric, and other similar baths used to strip metal.
Exhaust/vent scrubber wastes include a variety of solutions with characteristics similar to the process and plating wastes.

Industrial wastewater includes the large volumes of rinse waters used throughout the finish operations. These waters usually require treatment prior to discharge. Additional volumes of wastewaters requiring treatment are produced through cooling, steam condensate, boiler blowdown, and from exhaust scrubbers.

Common solid wastes generated by this industry include:

Industrial wastewater treatment sludge includes containing metals such as copper, chromium, nickel, tin, zinc, cadmium as well as the base metal of the substrate
Miscellaneous solid wastes include absorbents, filters, and empty containers

4. Wastewater Treatment

4.1. Discharge Regulatory History. Metal finish discharges are subject to environmental regulations depending upon the location of the discharge. Discharges to a receiving stream are regulated under the National Pollution Discharge Elimination System (NPDES) program under the federal Clean Water Act. The NPDES program is usually administered by the state agency. Discharges to a Publicly Owned Treatment Works (POTW) are subject to the effluent guidelines set in 40 CFR 433. These standards were established based on operational data from facilities utilizing conventional wastewater treatment methods. In many cases, federal limits do not meet the effluent limits of the local POTW. In these instances, local limits supersede federal standards and are utilized by the local POTW.

In establishing discharge standards, EPA was encouraging use of pollution prevention technologies for wastewater treatment from the industry. Conventional treatment could generally satisfy the discharge standards, but resulted in vast quantities of metal-bearing wastewater treatment sludge.

4.2. Conventional Industrial Wastewater Treatment. Figure 2 depicts conventional wastewater treatment systems in the metal finishing industry [8]. Treatment systems for cyanide plating which also uses chromium would use two separate pretreatment systems, followed by combined final treatment.

4.2.1. Chromium Treatment. Hexavalent chromium in baths and rinsewaters must be converted to the less toxic and easier to precipitate trivalent form. This chemical reaction is accomplished by use of reducing agents such as sodium metabisulfite, sulfur dioxide, or ferrous sulfate.

4.2.2. Cyanide Treatment. Cyanide baths and rinsewaters must be oxidized before discharge. Sodium hypochlorite is typically used for oxidation, but cyanide can also be destroyed by ozone or hydrogen peroxide treatment.

4.2.3. Combined Treatment. Once the chromium- and cyanide-bearing wastewaters have been treated separately, these acidic and alkaline waters are mixed together, along with metered amounts of spent process baths (such as nitric acid and cleaners). Final precipitation of the metals occurs in this step following comingling of the wastewaters. By adjusting the pH and using coagulants and polymers, precipitated particles flocculate, settle out, and eventually form metal hydroxide sludge. This sludge is then thickened, the water removed by physical means of water separation such as a filter press, and the wastewater treatment sludge (a listed hazardous waste for most metal finishers) is generated for final disposal.

TYPICAL ZINC PLATING LINE

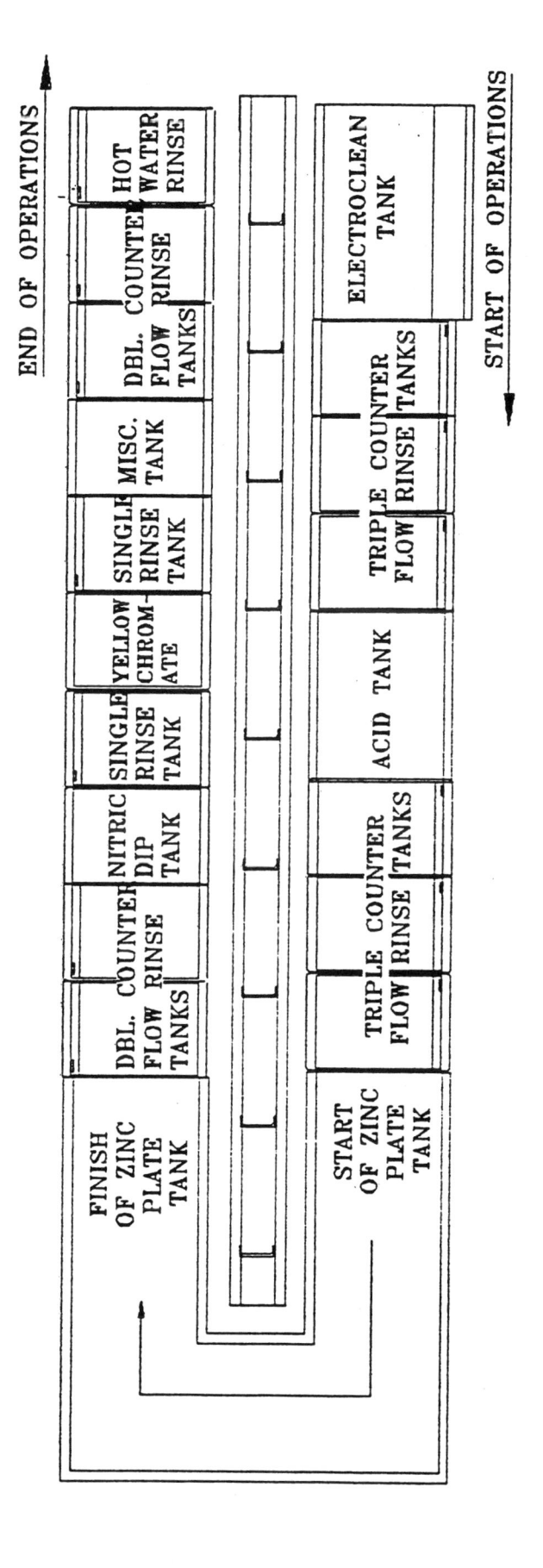

END OF OPERATIONS →

| FINISH OF ZINC PLATE TANK | DBL. COUNTER FLOW RINSE TANKS | NITRIC DIP TANK | SINGLE RINSE TANK | YELLOW CHROMATE | SINGLE RINSE TANK | MISC. TANK | DBL. COUNTER FLOW RINSE TANKS | HOT WATER RINSE |

| START OF ZINC PLATE TANK | TRIPLE COUNTER FLOW RINSE TANKS | ACID TANK | TRIPLE COUNTER FLOW RINSE TANKS | ELECTROCLEAN TANK |

START OF OPERATIONS →

FIGURE #1

CONVENTIONAL WASTE TREATMENT

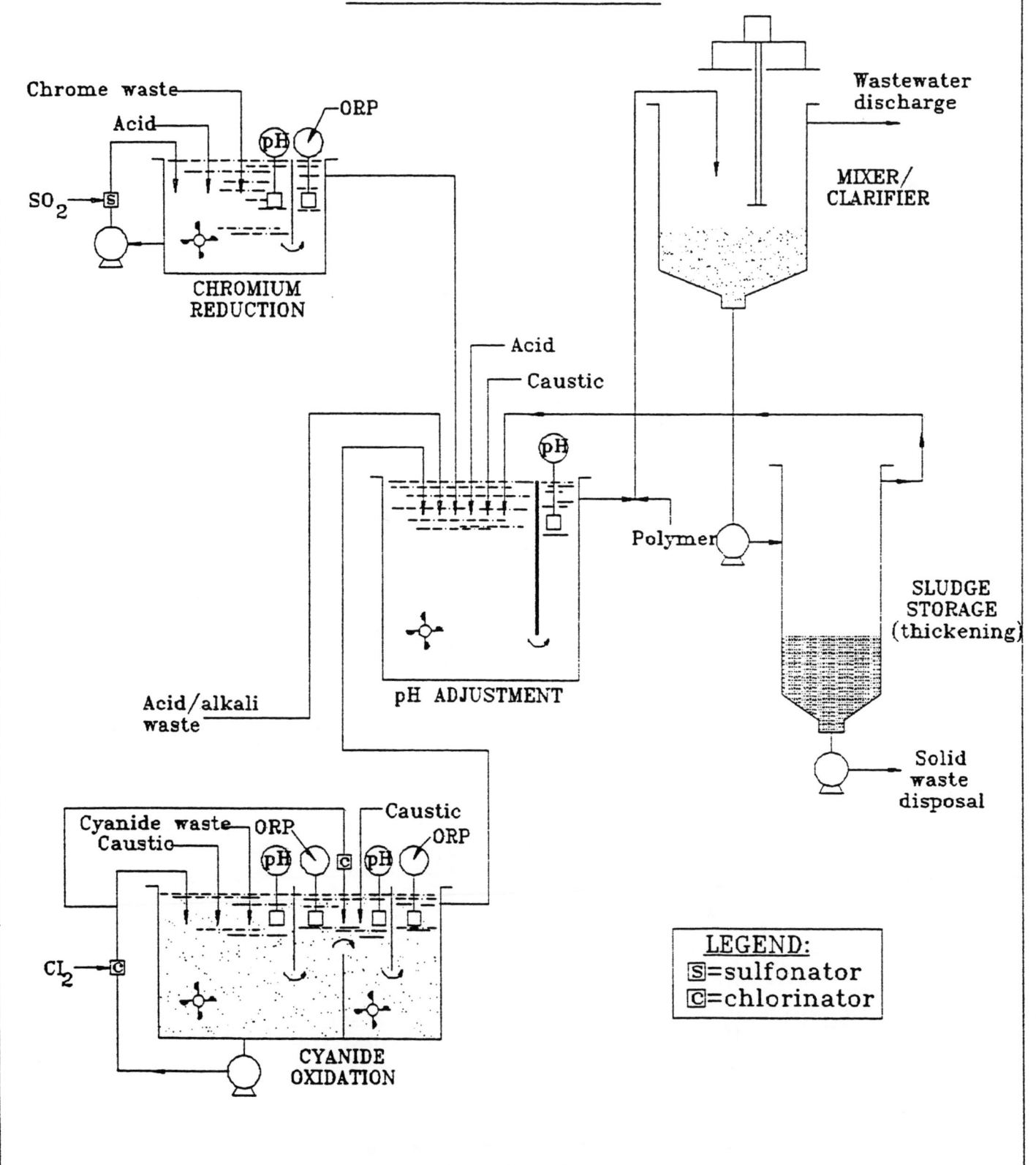

FIGURE #2

4.3. Downstream Disposal Minimization. A brief review of the variety of chemicals added in conventional treatment provide an understanding of the amount of sludge that can be generated. The typical treatment process includes: 1) the metals and salts from the process and plating baths, 2) salts and impurities that may have been brought into the process by the rinsewater, 3) the reducing or oxidizing chemicals (sodium metabisulfite and sodium hypochlorite, respectively), 4) sodium hydroxide or other neutralizing chemicals which adjust the pH to the optimal level required for metal precipitant formation and discharge compliance, 5) co-precipitants like magnesium hydroxide, ferrous sulfate, or lime that may have been added, and 6) organic polymers that may have been added to enhance precipitation.

With the addition of each chemical used in treatment, greater amounts of treatment sludge are generated for disposal. Because of costs associated with handling and disposal, wastewater treatment sludge can be the most expensive waste stream in the typical metal finishing operation.

Thus, any steps that can decrease the amount of metal hydroxide sludge are desirable. As a result, significant research has been done within the treatment industry to decrease the volume of sludge generated by conventional treatment systems. Because this work was performed downstream at the waste treatment system, only waste reduction was examined, not pollution prevention.

5. Upstream Pollution Prevention

Within the last decade, upstream approaches for pollution prevention, such as source reduction and recycling/resource recovery, have become the new industry focus.

Source reduction minimizes the quantities of wastes generated that will require treatment or disposal and is usually the least expensive method of waste reduction. Many source reduction options require only simple housekeeping changes or minor in-plant process modifications. Source reduction opportunities also exist for process baths and rinse systems.

With recycling and resource recovery, waste is used as raw material for the same or similar processes and valuable materials (such as the nickel in a nickel plating solution) are recovered from the spent solution prior to treatment or disposal.

The information in Secs. 6, 7, 8, and 9 concerning source reduction and recycling/resource recovery has been extracted from the EPA's *Guide to Pollution Prevention in the Metal Finishing Industry* [9] and can be viewed as state-of-the-art within this segment of the metal finishing industry.

6.0 Process Bath Source Reduction

Source reduction for the metal finishing industry at the process bath level can be achieved by material substitution that extends bath life and reduces drag-out.

6.1 **Material Substitution.** Increasingly stringent pollution control regulations continue to provide the incentive for less toxic process chemicals, and chemical manufacturers are gradually responding and introducing such substitutes. Elimination of process materials such as hexavalent chromium and cyanide-bearing cleaners and deoxidizers is obviously the preferred alternative, particularly since special equipment is needed to detoxify both.

Because there can be disadvantages in substituting one process chemical for another, the following key questions should be asked:

> Are substitutes available and practical?
> Will substitution solve one problem but create another?

Will tighter chemical controls be required of the bath?
Will product quality or production rate be affected?
Will the change involve any cost increases or decreases?

Most opportunities to reduce waste by substituting materials require modifying the chemistry of process baths or replacing the chemicals used for a particular process. Since process bath chemistries vary widely from plant to plant, these options are only described in general terms.

6.1.1. Purified Water. Deionized, distilled, or reverse osmosis water can be used instead of tap water for process bath makeup and rinsing. Natural contaminants, such as calcium, iron, magnesium, manganese, chlorine carbonates, and phosphates (found in tap water) reduce rinse water efficiency, interfere with drag-out recovery, and increase the frequency of process bath dumping [10]. These contaminants also contribute to sludge volume when they are removed from wastewater during treatment.

6.1.2. Hexavalent Chromium Alternatives. Trivalent chromium plating solutions can be used for decorative chromium plating in place of the more toxic hexavalent chromium. Although drag-out is decreased because trivalent chromium plating baths operate with a lower viscosity and lower concentration than hexavalent baths, the use of trivalent chromium eliminates the need to reduce the chromium before precipitation. In addition, using trivalent chromium eliminates the problems associated with hexavalent chromium bath misting and fugitive emissions in air scrubbers. Unfortunately, trivalent chromium is not presently available for hard chromium plating [11]. Other chromium alternatives include sulfuric acid and hydrogen peroxide (for chromic acid pickles, deoxidizers, and bright dips) and benzotriazole (0.1 to 1.0 percent solutions in methanol), or water-based proprietaries (for chromium based anti-tarnish). However, the latter two alternatives are extremely reactive and require ventilation.

6.1.3. Nonchelated Process Chemicals. Chelators are used in chemical process baths to control the concentration of free metal ions in the solutions. They are usually found in baths used for metal etching, cleaning, and electroless plating. Because chelating compounds inhibit the precipitation of metals, additional treatment chemicals must be used that may increase the volume and/or toxicity of the hazardous waste sludge. For example, when ferrous sulfate, a popular precipitant, is used to precipitate metals from chelated complexes, the precipitant adds significantly to sludge volume. In some applications, ferrous sulfate is added in large amounts at an 8-to-1 ratio to the contaminant metals [12]. Spent process baths containing chelators that cannot be treated on site must be transported off-site for treatment or disposal resulting in increased costs.

Several chelators are used in metal finishing industry processes. In general, mild chelators such as phosphates and silicates are used for cleaning and etching processes, while stronger chelating compounds (citric acid, maleic acid, and oxalic acid) are used in electroless plating baths. Ethylenediaminetetraacetic acid (EDTA) is also used but with less frequency than others [13]. While chelators help extend bath life, chelated process chemicals in wastewater must be removed to in compliance with discharge levels. Often, the pH of waste streams must be adjusted to break down the metal complexes that chelators form. EDTA, for example, requires lowering the pH below 3.0 to break the metal complex and allow subsequent metal precipitation at high pH [14].

Nonchelated process chemistries can be used in processes such as alkaline cleaning and etching where metals removed from workpiece surfaces do not need to remain in solution. In these cases, the metals can precipitate, and the process bath can be filtered to remove the solids. Note, however, that for electroless plating, it is less feasible to use nonchelated chemistries because the chelators allow the plating bath to function [13].

Nonchelated process cleaning baths usually require continuous filtration to remove precipitated solids. These systems generally have a 1-to-5 micron filter with a pump that can filter the tank's contents once or twice each hour [14].

Benefits of the nonchelated processes include: 1) reduced waste treatment and sludge handling and disposal costs for spent baths, 2) the metal-removal procedure during wastewater treatment is usually improved, and 3) the treated effluent is more likely to meet discharge requirements.

6.1.4. Noncyanide Process Chemicals. An alkaline chlorination process requiring sodium hypochlorite or chlorine is typically used to treat waste streams containing free cyanide. If complex cyanides are to be treated, ferrous sulfate precipitation is commonly used. The addition of these chemicals contribute to sludge volume. Although use of noncyanide process chemistries eliminate a treatment step, many noncyanide processes are difficult to use and can produce more sludge than cyanide baths. As with any change, the user should weigh the advantages and disadvantages for specific applications.

The potential wastewater treatment cost savings will depend on the cyanide treatment method and the volume of waste. Cyanide is typically oxidized with sodium or calcium hypochlorite. The use of noncyanide plating baths could eliminate or reduce this cost.

Alternatives to cyanide cleaners include trisodiumphosphate or ammonia; both provide good degreasing when used hot in an ultrasonic bath. However, they are highly basic and may complex with soluble metals if used as an intermediate rinse between plating baths.

Noncyanide alternatives to cyanide plating bath chemistries are also available but may be more costly then conventional cyanide baths. Acid tin chloride, for example, works faster and better than tin cyanide. Copper sulfate baths which are highly conductive and have a simple chemistry may be substituted for copper cyanide plating baths. Sulfate baths are economical to prepare, operate, and treat. Previous sulfate bath problems have been overcome with new formulations and additives [15]. The copper cyanide strike may still be needed for steel, zinc, or tin-lead base metals.

6.1.5. Alkaline Cleaners. A variety of chlorinated and nonchlorinated solvents are used to degrease workpieces before processing. Typically, these solvents are either recycled on site creating a solvent sludge that must be disposed of or transported off site for recycling or disposal. In contrast, use of hot alkaline cleaning baths in place of solvents permits on-site treatment and discharge to POTWs [11]. In the process, less sludge is generated than by solvent degreasing. The effectiveness of alkaline cleaners can be enhanced by applying an electrocurrent, a periodic reverse current, or ultrasonics. The benefit of solvent vapors and sludge elimination often outweigh any additional operating costs.

6.1.6. Alternative Cleaners. Although biodegradable cleaners may be acceptable for discharge to public sewers, the increased oxygen demand during treatment and disposal of the bath may slightly raise sewer fees. Nonphosphate cleaners may reduce waste by eliminating phosphate sludges during wastewater treatment. These and other alternative cleaners should first be tested to determine their effectiveness.

6.2. Extending Bath Life. Disposal of spent baths involves either on-site treatment or off-site disposal. Extending bath life, therefore, results in cost savings. Waste volume and bath replacement costs can be decreased through a variety of methods including filtration, replenishment, electrolytic dummying (i.e., using a low current to plate out contaminants), precipitation, monitoring, housekeeping, drag-in reduction, purer anodes and bags, and ventilation/exhaust systems. Each of these methods is described below.

6.2.1. Filtration. Filtration systems remove accumulated solids that reduce the effectiveness of the process bath operations. Continuous filtration of the bath removes these solids, thereby extending the life of the bath. Many acidic electroplating baths (e.g., acid copper sulfate, acid zinc, nickel sulfonate, nickel chloride) are currently filtered to improve quality. For other electroplating baths, filtration may not extend life significantly. Replacement of the filter media will generate a solid waste that adds to the operating costs. However, some filters utilize a reusable filter media which may help to alleviate filter element disposal costs.

6.2.2. Replenishment. Because the effectiveness of a cleaning bath decreases with use, it must be replaced or replenished. For a time, chemicals and water can be added to existing baths to extend their life. Early in the bath life, chemical replenishment reduces drag-out. However, over time the concentration of contaminants in the bath increases to the point where it becomes more expensive to add chemicals than to replace the entire bath with a new solution. At this point, the bath should be replaced. Although this approach does not ultimately reduce drag-out, it is still justifiable on the basis of quality control and waste reduction.

Various automated bath monitoring and replenishing systems are now available to extend bath life. Using data generated by bath monitoring systems, operators manually adjust and maintain process bath characteristics, such as pH, chemical concentration, and metal content, to improve product quality and extend bath life. Automated systems are also available to replenish and adjust the bath.

6.2.3. Electrolytic Dummying. Metal contaminants (such as copper) introduced into plating baths with workpieces degrade the effectiveness of the plating process. In zinc and nickel baths, copper can be removed by a process called "dummying." The process is based on the electrolytic principle that copper can be plated at a low electrical current. When the copper content becomes too high, an electrolytic panel is placed in the process bath. A "trickle current" is run through the system, usually at a density of 1 to 2 amperes per square foot. At this current, the copper in the bath solution plates-out on the panel, but the plating bath additives (such as brighteners) are unaffected. While some of the plating metals (zinc, nickel) are inadvertently removed, the savings realized by extending bath life justifies the slight metal loss.

6.2.4. Precipitation. Metals such as lead and cadmium enter the bath as impurities in anodes and can be removed from certain plating baths by precipitation. For a zinc cyanide bath, zinc sulfide can be added to precipitate lead and cadmium, and the precipitant can then be removed by filtration. As with all chemical reactions, care must be taken to ensure that precipitating reagents are compatible with bath constituents. In addition, iron and chromium contamination is common in acidic nickel baths. In most solution formulations, these metals can be removed using peroxide, pH elevation, and bath filtration.

6.2.5. Monitoring. Careful monitoring is key to determining the need to add chemicals or remove contaminants. The life of process bath can be extended only through continuous analysis of bath parameters, e.g., pH and metal content. In addition, a thorough understanding of the effect of contaminants on the production process is critical to reducing waste volume and the number of rejected parts that must be stripped and replated. Monitoring must be treated as an ongoing process, not an event.

6.2.6. Housekeeping. Good housekeeping practices that prevent foreign material from entering or remaining in a bath prolongs its life. When a part falls off the rack into a bath, it should be removed to reduce contamination of the bath. The racks should also be kept clean and free of contaminating material. Other housekeeping measures include protecting anode bars from corrosion, using corrosion-resistant tanks and equipment, and filtering incoming air to reduce airborne contaminants.

6.2.7. Drag-In Reduction. Drag-in resulting from liquids clinging to workpieces from preceding baths can shorten and reduce the effectiveness of subsequent baths. Rinsing to remove residual liquids (drag-out) prevents cross-contamination between baths.

6.2.8. Purer Anodes and Bags. Impurities contained in less pure anodes contaminate process baths; pure anodes do not contribute to bath contamination, but may cost more. Cloth bags around anodes prevent insoluble impurities from entering a bath but must be maintained and compatible with the process solution.

6.2.9. Ventilation/Exhaust Systems. Scrubbers, demisters, and condensate traps remove droplets and vapors from the air passing through ventilation and exhaust systems. If segregated, some wastes from scrubbers can be returned to process baths after filtering. Updraft ventilation allows mist to be collected in the ductwork and flow back to the process tank. For example, hard chromium plating baths would benefit from an updraft ventilation system.

Process baths that generate mist (e.g. hexavalent chromium plating baths, air agitated nickel/copper baths, etc.) should be in tanks with more freeboard to reduce the amount of mist reaching the ventilation system. This is, the added space at the top of the tank allows the mist to return to the bath before it is entrapped with the air entering the exhaust system. Foam blankets or floating polypropylene balls can also be used in hard or decorative chromium baths to keep mists from reaching the exhaust system.

6.3. Minimizing Drag-Out. Several factors contribute to drag-out including workpiece size and shape, viscosity and chemical concentration, surface tension, and temperature of the process solution [16]. By reducing the volume of drag-out that enters the rinse water system, valuable process chemicals can be prevented from reaching the rinse water, thereby reducing sludge generation. The techniques available to reduce process chemical drag-out include:

> Minimizing bath chemical concentrations by maintaining chemistry at the lower end of operating range.
> Maximizing bath operating temperature to lower the solution viscosity.
> Using wetting agents in the process bath to reduce the surface tension of the solution.
> Maintaining racking orientations to achieve the best draining.
> Withdrawing workpieces at slower rates and allowing sufficient solution draining before rinsing.
> Using air knives above process tanks.
> Using a spray or fog rinse above process tanks.
> Avoiding plating bath contamination.
> Using drain boards between process and rinse tanks to route drippage back to process tanks.
> Using drag-out tanks to recover chemicals for reuse in process baths.

A few of these drag-out reduction techniques require little if any capital investment; however, they do require training. For example, removing workpiece racks at a slower rate or allowing the rack to drain over the process tank for a longer time requires a conscientious operator. These procedures should not significantly affect production and should result in reducing process chemical purchases, water and sewer use fees, treatment chemical purchases, and sludge handling/disposal costs.

Other drag-out techniques require some capital expenditure. Drip bars can be installed above hand-operated process tanks to allow drag-out from workpiece racks to drain back into the process tank. If PVC piping is used and installation is performed by plant personnel, this option should cost no more than a few hundred dollars for five to eight tanks. Each of the suggested drag-out minimization techniques used is described below.

6.3.1. Process Bath Operating Concentration. Drag-out can be reduced by keeping the chemical concentration of the process bath at the lowest acceptable operating levels. Generally the greater the concentration of chemicals in a solution, the greater the viscosity [16]. As a result, the film that adheres to the workpiece as it is removed from the process bath is thicker and will not drain back into the process bath as quickly. This phenomenon increases the volume as well as the chemical concentration of the drag-out solution.

Chemical product manufacturers may recommend an operating concentration that is higher than necessary. Metal finishers should therefore determine the lowest process bath concentration that will provide adequate product quality. This can be accomplished by mixing a new process bath at a slightly lower concentration than is normally used. As the process bath is replenished, the chemical concentration can be reduced until product quality is affected. At this point, the process bath that provides adequate product quality at the lowest possible chemical concentration is identified. Alternatively, the new bath can be mixed as a low concentration, and the concentration can be gradually increased until the bath adequately cleans, etches, or plates the test workpieces. Fresh process baths can often be operated at lower concentrations than used baths. Makeup chemicals can be added to the used bath to gradually increase the concentration to maintain effective operation.

6.3.2. Process Bath Operating Temperature. Higher temperature baths reduce the viscosity of the process solution which enables the chemical solution to drain from the workpiece faster, thereby reducing drag-out loss. However, very high temperatures should be avoided because brighteners break down in most plating solutions and, in cyanide solutions, carbonate buildup increases. High temperatures may also cause the process solution to dry onto the workpiece as it is removed, increasing drag-out. Operating process baths at higher temperatures will also increase the evaporation rate from the process tank. To retain some of the advantages of higher temperature baths, water or process solution from a rinse tank can be added to replenish the process bath and to maintain the proper chemical equilibrium. Deionized water should be used to minimize natural contaminant buildup (such as calcium, iron, magnesium, carbonates and phosphates) in the process bath.

6.3.3. Wetting Agents. The addition of wetting agents to a process bath reduces the surface tension of a solution and, as a result, reduces drag-out loss by as much as 50 percent [16]. However, wetting agents can create foaming problems in process baths and may not be compatible with waste treatment systems. For these reasons, impacts on the process bath and the treatment system should be evaluated before using wetting agents.

6.3.4. Workpiece Positioning. Drag-out loss can be reduced by properly positioning workpieces on the rack. Workpieces should be oriented so that chemical solutions can drain freely and not get trapped in grooves or cavities. Following are suggestions for orienting and positioning workpieces:

> Parts should be tilted so that drainage is consolidated. The part should be twisted or turned so that fluid will flow together and off the part by the quickest route.

Avoid, where possible, positioning parts directly over one another.
Tip parts to avoid table-like surfaces and pockets where solution will be trapped.
Position parts so that only a small surface area comes in contact with the solution surface as it is removed from the process bath [16].

6.3.5. Withdrawal and Drain Time. The faster an item is withdrawn from the process bath, the thicker the film on the workpiece surface and the greater the drag-out volume. The effect is so significant that it is believed that most of the time allowed for draining a rack should instead be used for withdrawal only [16]. At plants that operate automatic hoist lines, personnel should adjust the hoist to remove the workpiece racks at the slowest possible rate. However, when workpieces are removed from a process bath manually, it is difficult to control the speed at which they are withdrawn. Nevertheless, supervisors and managers should emphasize to process line operators that workpieces should be withdrawn slowly.

The time allowed for draining can be inadequate if the operator is rushed to remove the workpiece rack from the process bath and place it in the rinse tank. However, a bar or rail above the process tank may help ensure adequate drain time prior to rinsing. If drip bars are used, employees can work on more than one process line or handle more than one rack during operation. This practice, "rotation plating," allows an operator to remove a rack from a plating bath and let it drain above the process tank while other racks are handled. Although increased drain time can have some negative effects due to drying, some baths (such as cleaners) are not affected. The operator can return after draining is completed to begin the rinsing stage.

6.3.6. Air Knives. Air knives can be used above process tanks to improve draining. As the workpiece rack is raised from the process tank, air is blown onto the surface of the workpieces to improve drag-out solution draining into the process bath. High humidity air can counteract workpiece drying.

6.3.7. Spray or Fog Rinses. Spray or fog rinse systems can be used above heated baths to recover drag-out solutions. If the spray rinse flow rate can be adjusted to equal the evaporation loss rate, the spray rinse solution can be used to replenish the process bath. Purified water should be used for the spray systems when possible to reduce the possibility of contamination entering the bath with the spray rinse water.

6.3.8. Plating Baths. Contaminated plating baths (for example, a cyanide plating bath contaminated with carbonate) increase drag-out by as much as 50 percent because of the increase in solution viscosity. Excess impurities also make application of recovery techniques difficult, if not impractical. Therefore, efforts should be made to reduce the level of impurities in the bath (e.g. by carbonate removal in cyanide baths).

6.3.9. Drain Boards. Drain boards capture process chemicals that drip from the workpiece rack as it is moved from the process bath to the rinse system. The board is mounted at an angle that allows the chemical solution to drain back into the process bath. Drain boards should cover the space between the process bath tank and the rinse tank to prevent chemical solutions from dripping onto the floor. Removable drain boards are desirable because they permit access to plumbing and pumps between tanks.

6.3.10. Drag-Out Tanks (dead or static rinse tanks). Process chemicals that adhere to the workpiece can be captured in drag-out tanks and returned to the process bath. Drag-out tanks are essentially rinse tanks that operate without a continuous flow of feed water. The workpiece is placed in the drag-out tank before the standard rinsing operation. Chemical concentrations in the drag-out tanks increase as workpieces are passed through. Since there is no feed water flow to agitate the rinse water, air agitation is often used to enhance rinsing. Eventually, the chemical concentration of the drag-out tank solution will increase to the point where it can be used to replenish the process bath. Drag-out tanks are primarily used with process baths that operate at an elevated temperature. Adding the drag-out tank solution back to the process bath compensates for evaporative losses resulting from high temperatures.

Deionized water should be used for drag-out tanks so that natural contaminants in tap water do not contaminate process baths when drag-out solutions are used to replenish them. Tanks should not be used to rinse workpieces from other process lines as contamination will result. Further, adding drag-out solution to some process bath chemistries (for example, electroless copper baths) can adversely affect the bath [17]. Often, a pretreatment step is required to remove contaminants prior to adding the recovered drag-out solution back to the process bath.

Generally, a drag-out tank can reduce both rinse water use and chemical loss by 50 percent or more [16]. To illustrate this savings, assume that a chemical bath loses approximately 2 gallons of drag-out each day, the total volume of drag-out loss each month would be 40 gallons, based on 20 work days per month. If the rinse system following the process bath operates at a flow rate of 5 gallons per minute for a total of 4 hours each day, water usage would be 24,000 gallons per month based on 20 work days per month. The savings in operation expenses include: (1) raw materials/chemical purchases, (2) water and sewer fees, and (3) treatment chemicals and sludge disposal. Reducing drag-out and rinse water use by 50 percent would reduce chemical losses by 20 gallons per month and water usage by 12,000 gallons if rinse water reduction is proportional to drag-out reduction. Sludge reduction and raw material/chemical reduction would increase savings significantly. The solution collected in the drag-out tank must be returned to the process bath when the concentration of the solution reaches the correct level. If it is returned at too low a concentration, the operating bath is diluted. If the concentration of chemicals in the drag-out tank gets too high (approaching bath concentration), the drag-out rinse becomes ineffective.

The overall savings for purchases of chemicals depend on the type of process chemicals and the amount of drag-out returned to the process tank. A corresponding savings in the purchase of treatment chemicals would also be realized by reducing rinse water effluent. Reducing the amount of sludge requiring off-site disposal will add to the savings.

The cost of a drag-out tank depends on the size of the tank. Since these tanks are not used as flow-through tanks, they can be set up without any plumbing. Typically, drag-out solutions are added back to the process bath manually, but automated systems maintain the best concentration in the drag-out tank and are more efficient. Technologies available to recycle process chemicals from drag-out tanks and rinse water effluent are discussed under Sec. 9, Recycling and Resource Recovery.

7.0   Rinse Systems Source Reduction

Most hazardous waste from a metal finishing plant is generated during the rinsing step that follows cleaning, plating, and stripping operations. Thus, aggressive source reduction practices with the rinse systems provide significant pollution prevention opportunities as well as savings from reduced water, sewer, and sludge disposal fees. By increasing rinse efficiency, a process line can reduce wastewater flow by as much as 90 percent [18, 19]. Improved rinse efficiency should also reduce treatment chemical use and sludge generation. These reductions are dependent on rinse water hardness and the sludge precipitation chemicals used in the wastewater treatment system.

Drag-out is the most significant source of process chemical loss. Treating rinse water containing these process chemicals generates hazardous sludge. The volume of sludge is proportional to the level of contamination in the spent rinse water.

By reducing the volume of rinse water containing process chemicals, resultant sludge generation will be reduced even if the total weight of the process chemicals remains constant. Two techniques available for reducing rinse water volume are improved rinse efficiency and rinse water flow control.

7.1. Improved Rinse Efficiency. The following three strategies can be used to enhance rinsing between various process bath operations: 1) turbulence between the workpiece and the rinse water, 2) increased contact time between the workpiece and the rinse water, and 3) increased volume of water during contact time to reduce the concentration of chemicals rinsed from the workpiece surface [16]. The third strategy, however, requires finishers to use significantly more rinse water than is actually necessary. Spray rinsing, agitation, increased contact time, rinse elimination, and counter-flow multiple tank rinsing, on the other hand, can be used to improve the efficiency of a rinsing systems and reduce the volume of rinse water.

7.1.1. Spray Rinses and Rinse Water Agitation. Turbulence, which involves spray rinsing and rinse water agitation, improves rinse efficiency. Although spray rinsing uses between one-eighth and one-fourth the volume of water of a dip rinse, it may not reach many parts of the workpiece [16]. To improve its effectiveness, however, spray rinsing can be combined with immersion rinsing. This technique uses a spray rinse as the first rinse step after the workpieces are removed from the process tank. A spray rinse removes much of the drag-out and returns it to the process bath before the workpiece is submerged into the dip rinse tank, permitting lower water flows in the rinse tank.

Spray or fog rinses can be installed above heated process tanks if the volume of rinse water from the spray system is less than or equal to the volume of water lost to heat evaporation. This practice allows the drag-out and the rinse solution to drain directly back into the process bath; in this way, the rinse solution replenishes the process bath. Deionized or reverse osmosis water should be used in this type of spray rinse system.

Workpieces can be agitated in the rinse water by moving the workpiece rack or creating turbulence in the water. Since most metal finishing plants operate hand rack lines, operators could easily move workpieces manually by agitating the hand rack. Rinsing is more effective if the pieces are raised and lowered into and out of the rinse tank rather than agitating the pieces while they are submerged.

The rinse water can also be agitated with forced air or water by pumping either air or water into the immersion rinse tank. Air bubbles create the best turbulence for removing the chemical process solution from the workpiece surface [16], but misting, as the air bubbles break the surface, may cause air pollution. Filtered air can be pumped into the bottom of the tank through a pipe distributor (air sparger) to agitate the rinse water. An in-tank pump can also recirculate the rinse water in the tank (a process known as forced water agitation). An agitator (mixer) can be used in a rinse tank, but this requires extra room in the tank to prevent parts from touching the agitator blades.

7.1.2. Increased Contact Time. By setting up multiple tanks in series as a counter-current rinse system, water usage can be reduced and contact time between the workpiece and the rinse solution can be increased. Rotation plating also increases contact time by allowing operators to leave workpiece racks in the rinse tanks while they handle other racks.

7.1.3. Rinse Elimination. The rinse between a soak cleaner and an electrocleaner may be eliminated if the two baths are compatible.

7.1.4. Counter-Current Rinse Systems. Multiple rinse tanks can be used to significantly reduce the volume of rinse water used. A multistage counter-current rinse system uses up to 90 percent less rinse water than a conventional single-stage rinse system [12]. In a multistage counter-current rinse system, workpiece flow moves in a direction opposite to the rinse water flow. Water exiting the first tank (the last tank in which the workpiece is immersed) becomes the feed water to the second tank. This water then feeds the third tank, and so on for the number of tanks in the line. Figure 4 illustrates the use of a three-stage counter-current rinse system.

The effectiveness of this multistage system in reducing rinse water use is illustrated in the following example. A plant operates a process line where the drag-out rate is approximately 1.0 gallon per hour. This process bath is followed by a single-stage rinse tank requiring a dilution rate of 1,000 to 1 to maintain acceptable rinsing. Therefore, the flow rate through the rinse tank is 1,000 gal/hr. If a two-stage counter-current rinse system were used, a rinse water flow rate of only 30 to 35 gal/hr would be needed. If a three-stage counter-current rinse system were used, only 8 to 12 gal/hr would be required [18].

A multistage counter-current rinse system allows greater contact time between the workpiece and the rinse water. Greater diffusion of process chemicals into the rinse solution results as more rinse water comes into contact with the workpiece. The disadvantage of multistage counter-current rinsing is the need for additional tanks and work space. Since many metal finishers lack room to install additional rinse tanks, multistage rinse systems are not always feasible. One option available to a metal finishing plants that lack floor space is to reduce the size of the rinse tanks or to segregate existing tanks into multiple compartments. This option is, of course, limited by the size of the workpieces.

7.2. Flow Controls. Oversized water pipes or continuous water flow when the rinse tanks are not in use results in excessive volumes of rinse water. Proper rinse water control devices can increase the efficiency of a rinse water system.

The cost of reducing rinse water use varies depending on the method. The cost may be limited to that associated with purchasing and installing flow restrictors or timers. Savings from reduced rinse water flow rates include direct reduction of water use, sewer fees, treatment chemical use purchases, and sludge generation.

7.2.1. Flow Restrictors. Flow restrictors limit the volume of rinse water flowing through a rinse system based on the optimal flow rate. Since most small- and medium-sized metal finishers operate batch process lines in which rinse systems are manually activated at the start and finish of operations, pressure-activated flow control devices, such as foot pedal-activated valves or timers, can ensure that water is not left on after the rinse operation is completed.

Installation of a flow restrictor upstream of all the rinse water influent lines also reduces water use. The flow restrictor should be set at a rate less than the flow rate required to operate all rinse tanks simultaneously. This setting requires operators to turn off the water in the unused rinse systems so the rinse systems in use will have adequate flow. For example, if a metal finisher operates between 20 and 24 separate rinse systems, each requiring an average flow rate of 2 gallons per minute (GPM), a flow restrictor, installed upstream of all the rinse water influent lines, could limit total water flow to 15 GPM. Therefore, operators must first turn off the unused rinse systems to ensure that the rinse systems requiring immediate use will operate properly. Operator training and cooperation are required, or parts will not be rinsed effectively, and product quality will decrease.

7.2.2. Conductivity-Actuated Flow Controller. A conductivity-actuated flow controller governs fresh water flow through a rinse system be means of a conductivity sensor that measures the level of ions in the rinse water. When the ion level reaches a preset minimum, the sensor activates a valve that shuts off the flow of fresh water into the rinse system. When the concentration builds to the preset maximum level, the sensor again activates a valve that opens to resume the flow of fresh water.

8. Housekeeping Source Reduction

Although the contribution of improved housekeeping to overall waste minimization if difficult to quantify, often simple housekeeping improvements can provide low to no cost opportunities for waste reduction. A plant can implement a number of housekeeping practices to reduce waste including: 1) developing inspection and maintenance schedules, 2) controlling the purchasing and handling of raw materials, 3) removing dropped parts quickly from baths, 4) keeping filters and other process equipment in good working condition, and 5) authorizing a limited number of employees to accept and test samples from chemical suppliers.

8.1. Inspection and Maintenance. Production, storage, and waste treatment facilities should be inspected regularly to identify leaks, improperly functioning equipment, and other items that may lead to waste. Frequent inspections can identify minor issues before they become significant problems. Equipment such as piping systems, filters, storage tanks, defective racks, air sparging systems, and automated flow controls should be maintained according to manufacturers' specifications and inspected on a regular basis. Operators' production procedures (such as drain time and rinse methods) should be reviewed to ensure compliance with company procedures and proper recordkeeping.

Dropped parts and tools should be removed from process baths quickly to reduce contamination of the bath. Easily accessible rakes to recover dropped items are essential. Maintenance schedules should be coordinated with inspection schedules to ensure that equipment is operating at optimal efficiency.

8.2. Chemical Purchasing and Handling. An excess inventory of chemicals as well as outdated chemicals and materials result in unnecessary and costly waste. Procedures to inventory and control the purchase of materials can prevent such waste. Opened containers should be completely empty before new containers are opened to reduce stockpiling of raw materials, the potential for spills, and the likelihood of mixing poor process baths.

In addition, strict procedures should be developed for mixing chemicals. Mixing procedures should be designed to minimize spills, provide correctly mixed booths, and ensure that the baths are operated at the lowest possible concentration to reduce drag-out loss. A limited number of trained personnel should be designated to mix chemicals. Such procedures will improve the consistency of the solution formulations and decrease waste.

8.3. Sample Testing. Many suppliers provide metal finishers with a variety of process chemicals for testing. However, unused material becomes waste and should not be allowed to stockpile at the site. If possible, metal finishers should accept test samples only if unused samples can be returned. The unused portion of analytical samples taken from process baths should be returned to the process bath.

9. Recycling and Resource Recovery

Recycling and resource recovery technologies either directly use waste from one process as raw material for another process or recover valuable materials from a waste stream before disposal. Some spent chemical process baths and the majority of rinse water can be reused in other plant processes. Process chemicals can be recovered from rinse water and sold or returned to process baths.

CLOSED LOOP SYSTEM

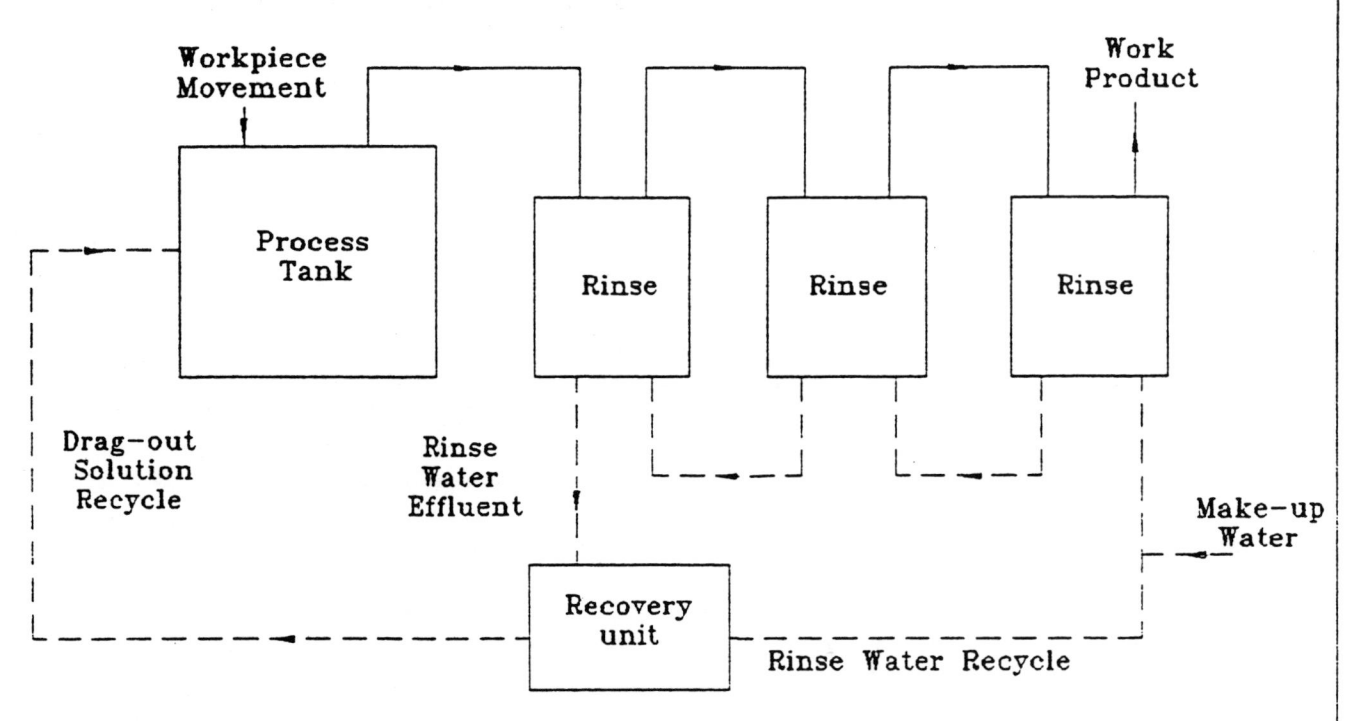

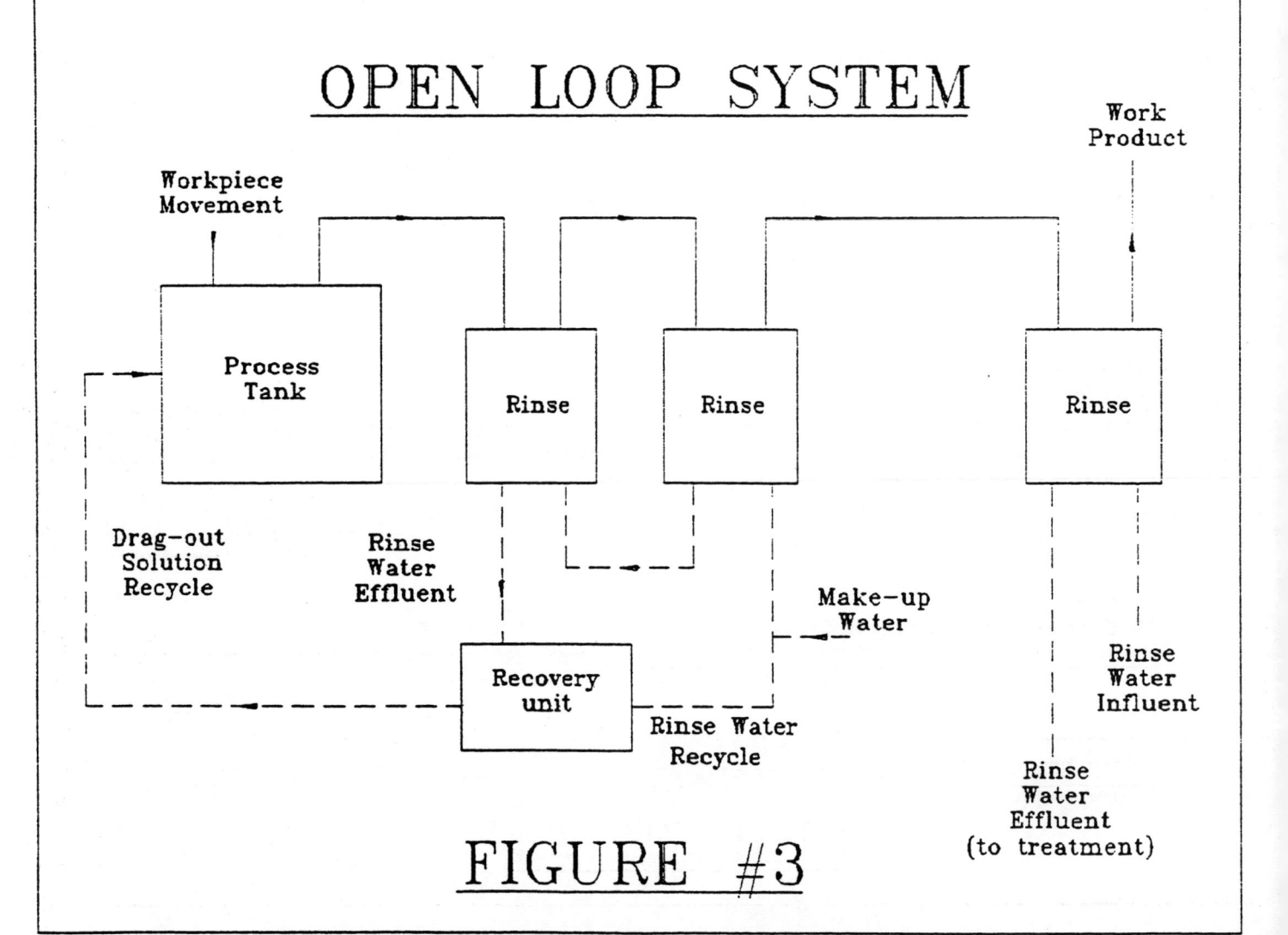

OPEN LOOP SYSTEM

FIGURE #3

ION EXCHANGE

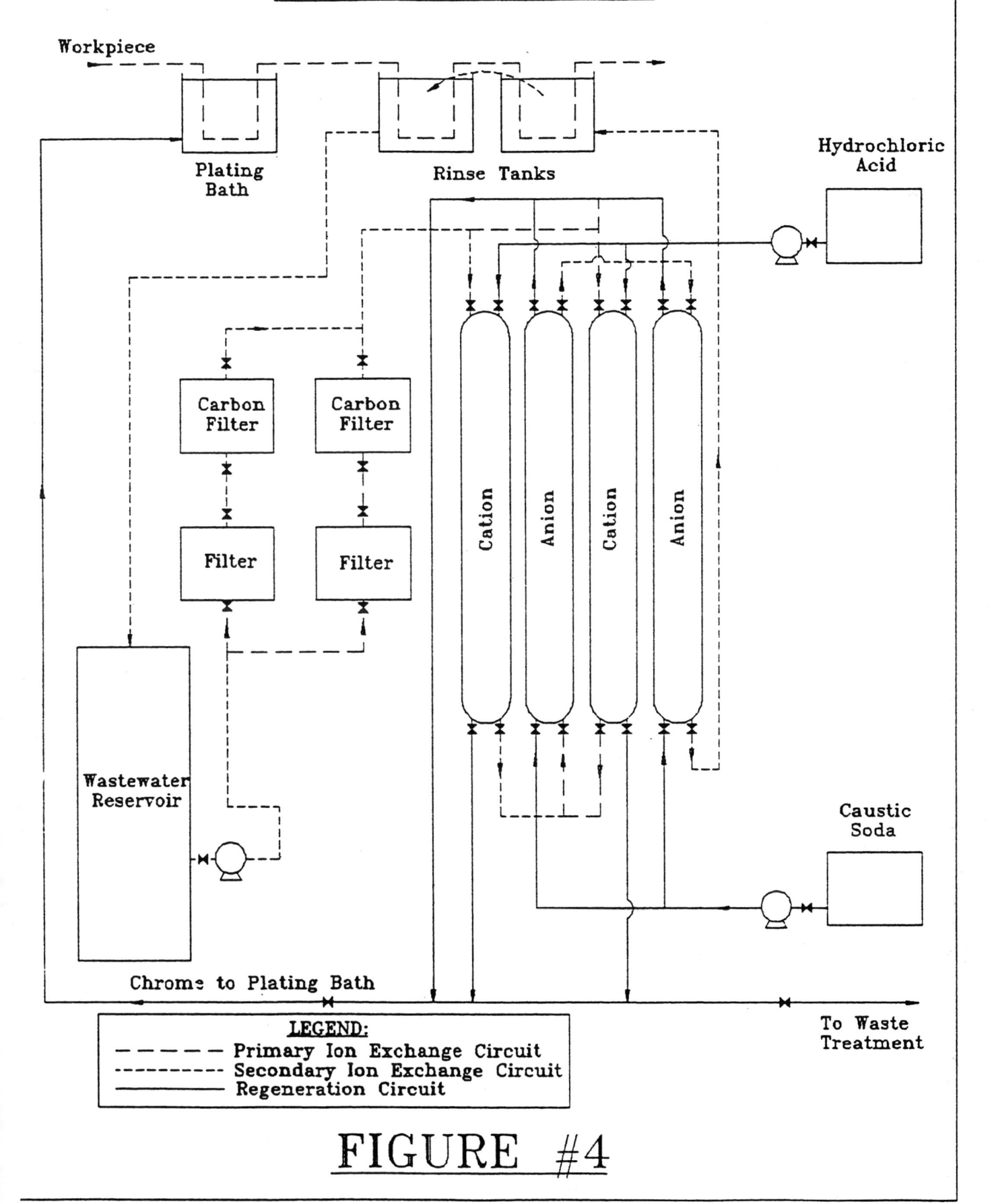

LEGEND:
- – – – Primary Ion Exchange Circuit
- - - - - Secondary Ion Exchange Circuit
- ——— Regeneration Circuit

FIGURE #4

Most recycling and resource recovery technologies require waste segregation before: 1) reuse of a waste material for another process, 2) recovery of valuable chemicals from a waste stream, or 3) recycling of rinse water. Therefore, recycling and resource recovery technologies typically will require process piping modifications and additional holding tanks to provide appropriate material segregation.

9.1. Reusing Waste Materials. The chemical properties of a waste stream must be understood to assess the potential for reusing the waste as a raw material. Although the chemical properties of a process bath or rinse water solution may make it unacceptable for its original use, the waste materials may still be valuable for other applications. Metal finishers should therefore evaluate waste streams for the properties that make them useful rather than the properties that render them waste.

9.1.1. Rinse Water. One waste material reuse option common among metal finishers is multiple-use rinse waters, in which the rinse water from one process is used for the rinse water of another. The primary cost associated with rinse water reuse is in replumbing the rinse system. Depending on the design of the rinse water reuse system, storage tanks and pumps may also be needed.

Rinse solutions too contaminated for their original purpose may be useful for other rinse processes. For example, effluent from a rinse system following an acid cleaning bath can sometimes be reused as influent to a rinse system following an alkaline cleaning bath. If both rinse systems require the same flow rate, 50 percent less rinse water would be used to operate them. In addition, reusing water in this way can improve rinse efficiency by accelerating the chemical diffusion process and reducing the viscosity of the alkaline drag-out film [16]. Care must be exercised to make sure that tank materials and pipes, as well as bath chemistries, are compatible with the rinse solutions.

Acid cleaning rinse water effluent can be used as rinse water for workpieces that have gone through a mild acid etch process. Effluent from a critical or final rinse operation, which is usually less contaminated than other rinse waters, can be used as influent for rinse operations that do not require high-rinse efficiencies. Another option is using the same rinse tank to rinse parts after both acid and alkaline baths. Metal finishers should evaluate the various rinse water requirements for their process lines and configure rinse systems to take advantage of rinse water reuse opportunities that do not affect product quality.

9.1.2. Spent Process Baths. Typically, spent acid or alkaline solutions are replaced when contaminants exceed an acceptable level. However, these solutions may remain sufficiently acidic or alkaline to act as pH adjusters. For example, alkaline solutions can be used to adjust the pH in a precipitation tank. Acid solutions can be used for pH adjustment in chromium reduction treatment. Since spent cleaners often contain high concentrations of metals, they should not be used for final pH adjustments. It is important to make sure the process solutions are compatible before mixing with other baths. In addition, chemical suppliers may have reclamation services which permit certain spent plating baths to be returned.

9.2. Recycling Rinsewater and Process Baths. Rinse water can be recycled in a closed-loop or open-loop system. In a closed-loop system, the treated effluent is returned to the rinse system realizing a significant reduction in water use and discharges to the wastewater treatment plant. A small amount of waste is still discharged from a closed-loop system. An open-loop system allows the treated effluent to be reused in the rinse system, but the final rinse is fed by fresh water to ensure high quality rinsing. Therefore, some treated effluent will continue to be discharged to the sanitary sewer. Figure 3 shows the configurations for both a closed-loop and open-loop rinse water recycling system.

To improve the economic feasibility of these systems, rinse water efficiency techniques should first be implemented. Multistage counter-current rinse systems, flow controls, and drag-out reduction techniques should be pursued to reduce the volume of water requiring treatment for recovery, thus reducing the equipment capital costs.

In the past, material recovery from metal finishing was not considered economical. However, effluent pretreatment regulations and treatment and disposal costs are now a significant economic factor. As a result, metal finishers may find it economical to reuse rinse water and to recover metals and metal salts from spent process baths and rinse water.

Recovered metal can be reused in three ways: (1) recovered metals (and process solutions) can be returned to baths as makeup, (2) metals can be sold or returned to suppliers, or (3) elemental metal can be sold to a reclaimer or reused on site as plating metal anode materials. Key technologies to successfully recover metals and metal salts include:

Evaporation
Reverse osmosis
Ion exchange
Electrolytic recovery (electrowinning)
Electrodialysis

Each of the above technologies could be generically represented by the recovery unit noted in Fig. 3. Also, these technologies can be used separately or in combination to recover chemicals from rinse water effluent.

The savings actually achieved through metal recovery will be site-specific and dependent on including the volume of waste that contains metals, the concentration of the metals, the potential to reuse some of the metal salts, and the treatment and disposal costs. Many systems may not be economically feasible for small metal finishers because the savings may not be great enough to achieve an acceptable payback on their investment.

9.2.1. Evaporation. Evaporation has been successfully used to recover a variety of plating bath chemicals. This simple technology is based on the physical separation of water from dissolved solids such as heavy metals. Water is evaporated from the collected rinse water to allow the chemical concentrate to be returned to the process bath. The drag-out recovered is often returned to the process tank in higher concentrations than in the original process solution. Water vapor is condensed and can be reused in the rinse system. The process is performed at low temperatures under a vacuum to prevent degradation of plating additives. Atmospheric pressure evaporators are used most commonly because of their lower capital cost. Evaporation is more economical when used with multistage counter-current rinse systems because the quantity of rinse water to be processed is small. The process is energy-intensive and becomes expensive for large volumes of water; heat pumps and multistage counter-current rinse systems have lower operation costs. Evaporation is most economical when the amount of water to be evaporated is small or when natural atmospheric evaporation can be used.

A variation on standard evaporation technology is the cold vaporization process, which works by a similar evaporation separation principle except that an increased vacuum evaporates water at temperatures of 50°F to 70°F. This type of evaporation system is less energy intensive than electrically heated systems because it utilizes heat from the air around the unit. Some equipment uses the heat generated from the vacuum system to provide the heat needed for evaporation.

9.2.2. Reverse Osmosis. Reverse osmosis (RO) is a pressure-driven process in which a semipermeable membrane permits the passage of purified water under pressure greater than the normal osmotic pressure, but does not allow larger molecular weight components to pass through. These concentrated components can be recovered and returned to the process bath, and the treated rinse water is then returned to the rinse system for reuse. The most common application of RO technologies in metal finishing operations is in the recovery of drag-out from acid nickel process bath rinses. Although the technology is designed to recover a concentrated drag-out solution, some materials (such as boric acid) cannot be fully recovered. Also, RO is a delicate process that is limited by the ability of the membranes to withstand pH extremes and long-term pressure. RO membranes are not generally suitable for solutions having high oxidation potential (such as chromic acid). Also, the membranes will not completely reject many nonionized organic compounds. Therefore, activated carbon treatment is typically required before the rinse water solution can be returned to the rinse system. Activated carbon can be costly, but for certain cases it may be the only practical approach.

9.2.3. Ion Exchange. Ion exchange (IX) can be used to recover drag-out from a dilute rinse solution. The chemical solution is passed through a series of resin beds that selectively remove cations and anions. As the rinse water is passed through a resin bed, the resin exchanges ions with the inorganic compounds in the rinse water. The metals are recovered by cleaning the resin with an acid or alkaline solution. The metals then can be electrowon from the resin regeneration solution while the IX treated water can be returned to the rinse system for reuse. IX units can be used effectively on dilute waste streams and are less delicate than RO systems, but the water must be filtered to remove oil, grease, and dirt to protect the resin. Certain other metals may eventually foul the resin, requiring a special procedure to remove the foulant.

Ion exchange is commonly used to treat rinse water from chromic acid process baths. Figure 4 shows how an IX unit can be used to recover chromic acid and reuse rinse water. The rinse water waste stream is filtered prior to passing through a cation column and two anion columns. The primary cation resin column removes heavy metals from the solution, while the anion resin column removes the chromate ions. The chromates are removed from the anion resin, with sodium hydroxide, forming sodium chromate which is then regenerated as chromic acid by passing through a secondary cation bed. The secondary bed replaces the sodium ions with hydrogen ions. The cation beds themselves are regenerated with hydrochloric acid and the spent regenerant solution is usually treated in the wastewater treatment system. It is important to note that chloride contaminates the chromium plating bath, and that treatment with silver nitrate to precipitate the chloride is expensive.

IX equipment requires careful operation and maintenance. In addition, recovery of chemicals from the resin columns generates significant volumes of regenerant and wash solutions, which may add to the wastewater treatment load.

9.2.4. Electrolytic Recovery/Electrowinning. Electrowinning is a process used to recover the metallic content of rinse water. It operates using a cathode and an anode, which are placed in the rinse solution. As current passes between them, metallic ions deposit on the cathode, generating a solid metallic slab that can be reclaimed or used as an anode in an electroplating tank. The electrowinning process is capable of recovering 90 to 95 percent of the available metals and has been successfully used to recover gold, silver, tin, copper, zinc, solder alloy, and cadmium [20].

Several basic design features well known to the electroplating industry are employed in electrolytic recovery including: 1) expanded cathode surface area, 2) close spacing between cathode and anode, and 3) recirculation of the rinse solution [21]. Electroplaters can design their own units by closely spacing parallel rows of anodes and cathodes in a plating tank and circulating rinse solutions through the tank. This process can also be used to recover metals from spent process baths prior to bath treatment in the wastewater treatment system.

High surface area electrowinning/electrorefining is another method of electrolytic recovery. The metal-containing solution is pumped through a carbon fiber cathode or conductive foam polymer, which is used as the plating surface [22]. To recover the metals, the carbon fiber cathode assembly is removed and placed in the electrorefiner which reverses the current and allows the metal to plate onto a stainless steel starter sheet. These systems recover a wide variety of metals and regenerates many types of solutions. High surface area metal recovery is used mainly with dilute solutions such as rinse water effluent.

9.2.5. Electrodialysis. Electrodialysis is used to concentrate and separate ionic components contained in rinse water solutions. A water solution is passed through a series of alternately placed cation-and-anion-permeable membranes. These membranes are placed parallel to the flow of water, and an anode and cathode are placed on opposite sides of the membrane stack. The anode and cathode create an electropotential across the stack of membranes, causing the ions in the rinse solution to migrate across the membrane. The selectivity of the alternating membranes causes both anions and cations to migrate into alternating channels, and ion-depleted water remains in the other channels. The concentrated solution can be returned to the plating baths, while the treated water is recycled through the rinse system.

10. Metal Finish Practicabilities

10.1. Myth of Simple Changes. In the previous sections, many current pollution prevention practices in the metal finishing industry were discussed. In many ways, these practices follow the law of motion - "For every action, there is an equal and opposite reaction." That is, before performing source reduction or recycling/resource recovery, potential changes or impacts on the rest of the metal finishing process must be considered. Though the resultant changes may not be totally equal, some change will result. For example, as mentioned in Sec. 6, material substitution may solve one problem but create another. Thus, the operation-wide impacts of any pollution prevention change must be considered, not just the benefits of a particular process. These additional considerations are detailed in Refs. [2] and [3].

Metal finishers, therefore, need to be cognizant of all the aspects of their operations so that they can objectively evaluate pollution prevention technology vendors claims. Vendors may state that drop in substitutes exist without the need to think about any other potential process issues or discharge issues. As with any purchase, ask the vendors hard questions and compile a detailed and complete list of the advantages and disadvantages [24].

10.2. Myth of Zero Discharge. Although much talk within the industry exists about "zero discharge," for any plating facility, zero discharge is not presently achievable. A proactive facility that employed all of the technologies and techniques presented in Secs. 6 through 9 would still have hazardous and non-hazardous wastes to handle and dispose. By-products from advanced separation technologies, such as reverse osmosis or electrodialysis (Secs. 9.2.2 and 9.2.5 respectively) still require either on-site neutralization, treatment, and discharge or off-site disposal. Reference [25] provides a good solid discussion about the myth of zero discharge and presents the results of a successful case study of a facility that implemented a closed-looped system for approximately 90 percent of its water.

Because in its true sense "zero discharge" is not presently attainable for the metal finishing industries, better terms might be "closed loop," "near zero discharge," or "zero wastewater discharge."

11. Future Areas of Potential

11.1. Solution Chemistry. Significant progress has been made in solution chemistry. As discussed in Sec. 6.1.4, non-cyanide replacements now exist for most metal cyanide plating solutions. Solution alternatives are still being sought for plating copper onto zinc die castings, brass, bronze and silver [23]. Extensive chemistry work should continue to find suitable non-cyanide alternatives that closely match the quality of cyanide plating for these remaining cyanide solutions. In many instances, cyanide plating solutions currently provide more consistent quality deposits compared to non-cyanide alternatives.

Significant advances in many of the uses of trivalent chromium to replace hexavalent chromium were discussed in Section 6.1.2. Even with these advances, presently no trivalent chromium solutions replace hexavalent chromium in hard chromium plating applications. Because hard chromium plating is still a substantial metal finishing market, research into innovative trivalent alternatives in hard chromium plating should continue. Likewise, research to find non-chromate sealers and conversion coatings which will have sufficient chromium-like properties need to continue. Most of the current non-chromate sealing and conversion coatings alternatives do not hold up to the same degree as chromate-containing ones in performance tests such as salt spray. Perhaps, current industry performance standards will need to be reevaluated to determine if they accurately reflect field performance requirements or need to be adjusted. Certainly, opportunities exist for companies to determine the actual required specifications so that alternative solutions, such as non-chromate sealers and conversion coatings, could be used in additional areas while decreasing the pollution from the process.

Other substitutions, such as zinc-alloy plating instead of cadmium, provide almost all of the cadmium plating benefits while providing significantly less toxic waste and wastewaters. New zinc alloys can yield salt spray resistance which matches or surpasses that of chromium but do not provide surfaces resistant to biological growth. Research to completely match plating characteristics of substitution solutions (such as zinc alloy) to the original (such as cadmium) should continue.

Opportunities will continually exist for advances in solution chemistry. These advances might involve combinations of non or lower toxic materials that perform like their toxic counterpart or generate less overall pollution. Opportunities also exist for further advances in technologies for more effective recycle or reuse. For example, enhanced recycling may become possible if chemical scavenging systems are improved to remove more impurities or if advances in chelating systems are made to permit easier precipitation.

11.2 Substrate Substitution. In many applications, metal products with decorative or protective finishes have already been replaced by non-plated parts or processes such as plastics. Substrate substitution was not discussed in Secs. 6 through 9 as these sections focused on pollution prevention methods in existing applications. A substrate substitution may be made for pollution prevention reasons, durability, improved function, or because parts which had received metal finishing were simply found not to require plating or finishing upon closer examination. While many substrate substitutions decrease the amount of metal finishing necessary and the waste generated, it is possible that pollution is transferred from one generator area (metal finishing) to another (such as plastics manufacturers). A complete and detailed life cycle analysis would be required to define the overall pollution reduction benefit.

11.3. Green Product Design. The total costs associated with waste generation and disposal as well as regulatory compliance have stimulated operational reviews throughout the metal finishing industry. These reviews are focusing on present processes and products as well as products presently under design.

More and more of the current manufacturing design processes include designing the product or its parts to minimize environmentally-related manufacturing costs. These new design processes may even allow for a "green" label in the future. By incorporating pollution prevention at a product's inception, manufacturing and marketing are being forced to work together, at a much earlier stage, to determine the exact needs of the component materials. In this process, they are finding that some metal finishing steps can be eliminated. Opportunities exist, therefore, for numerous players (consultants, laboratories, engineers of all sorts) to be involved with industry and product development to incorporate pollution prevention and the "greening" of products.

In the past, parts were never specifically designed with plating considerations in mind. Engineers would first perform the physical design, then examine the specific needs of the part, such as required corrosion protection, hardness, etc. Increasingly, parts are designed around the limitations of existing plating and treatment technologies. For example, new generations of parts are being designed to utilize technologies like vacuum deposition, especially when this technology is already in place at a facility for other products or available at a nearby contract facility.

To illustrate the need for an integrated approach in green product design, consider the following example. Manufacture of a green product would include: 1) minimizing required chemical operations at the start, 2) using alkaline cleaners (Sec. 6.1.5) instead of ozone depleting chemicals like 1,1,1 trichloroethane, 3) plating with non-cyanide solutions (Sec. 6.1.4), and 4) utilizing bath, rinsewater, housekeeping, and recycling source reductions activities (Secs. 6 through 9). The greening of products will require a systems approach to the product development and manufacturing life cycle.

11.4. Non-Chemical Cleaning. New twists on existing technologies may reduce or eliminate waste streams associated with metals cleaning prior to plating or painting. The use of media blasting, ice crystal blasting, and frozen carbon dioxide (dry ice) are now being used in limited production for stripping of organic coatings. Because cleaning and stripping action is put directly at the surface, these technologies may eliminate the need for wet chemical cleaning (or at least one of the required cleaning steps) in cleaning/stripping of metals and/or organic coatings.

11.5. Combinations of Technology. The careful and controlled use of a combination of technologies may provide further environmental pollution prevention opportunities. For example, mechanical stripping alone may not provide the results desired for organic coating or metal removal, but a combination of mechanical and chemical stripping may yield the desired quality and generate less waste. This joint approach may also enhance limited recycling of materials not possible with either technique separately.

Trends to package related pollution prevention technologies in an attempt to make pollution prevention "easier" are growing. One example is the skid-mounted packages that combine ion exchange with electrolytic technology for recycling and metals recovery from select plating baths. Source segregation and concentration of waste (as will be discussed in Secs. 11.7, 11.8, and 11.9), such as evaporation technology will likely be included more in these packaged units. Opportunities for technical review of individual process needs and design and installation the appropriate packages of equipment will likely grow.

11.6. Vacuum Deposition. Vacuum deposition has existed for many years and has a proven track record in certain industries. As compared to conventional metal finishing processes, vacuum deposition generates less pollution, is much less chemical intensive, and achieves the same desired surfaces. The effective use of vacuum deposition can eliminate the need for many or most of the pollution prevention opportunities discussed in Secs. 6 through 9. The opportunity certainly exists for increased application of this technology to a wider variety of parts and orientations. Breakthroughs will likely be achieved in reducing the capital equipment costs and in covering problematic parts more adequately and uniformly. Successful breakthroughs will likely enable capture of a larger portion of the potential metal finishing market.

11.7. Centralized Processing. Centralized processing of metal finishing wastes for reuse and recovery has been used for many years in other countries, most notably Japan and Europe.

This approach works best when the source materials from metal finishing are segregated and concentrated using evaporation or other techniques. The closer the wastes match the baths within the plating systems, the greater the chance of success with centralized processing. For this reason, reprocessing works only with concentrated solutions and not rinsewaters. If rinsewaters are to be reprocessed, a concentrating method such as ion exchange (Sec. 9.2.3) or one of the membrane technologies (Secs. 9.2.2 and 9.2.5) must first be applied. Further concentration of the waste through evaporation (Sec. 9.2.1) may also be required. Once waste streams and metals are combined, it is much more difficult, both technically and economically, to separate them during centralized reprocessing.

One variation of centralized processing is presently available for metal finishing wastes, but it does not reuse and recover materials. Facilities utilizing this method generally take spent plating baths or their associated waste treatment sludges and render them nonhazardous waste products which do not undergo any recovery of metals or chemicals. Other limited facilities, such as secondary metal smelters, actually take select metal finishing wastes and sludges and recover metals such as nickel and chromium for use in the iron and steel industry.

The utopian processing facility would take segregated metal finishing wastes and, by using environmentally sound practices, separate and recover the metals via hydrometallurgical processes. Again, segregated wastes are required. Perhaps niche market facilities could be strategically located near large geographic plating centers for particular plated metals. In this way, each separate plating facility under contract would not need to purchase recovery equipment on its own; rather, it would utilize the central facility.

11.8. Metal Hydroxide Management. Thousands of metal finishing treatment systems across the country, generate metal hydroxide sludge or filter cake as the typical end products. Many of these facilities operated for years before environmental regulations, and a substantial amount of metal hydroxide waste is buried in the ground, either at company-owned monofills, or commercial and municipal landfills. At present, only limited facilities (secondary smelters) can reclaim metals from limited quantities of high value/highly segregated treatment sludges. Research and development into metal reuse and reclaim of metal hydroxides could result in reclaim of metals, formation of another product, or reusable by-products. Industrial monofills, particularly those dedicated to one consistent process line, present potential opportunities as well. For example, a metal hydroxide monofill from an operation using higher value metals might contain sufficient concentrations to warrant recovery investigation.

11.9. Vendor Processing of Spent Solutions. As with any centralized processing, vendor processing yields optimal results with waste segregation. Rinsewaters are not amenable to this vendor processing as discussed in Sec. 11.7. Concentrating techniques will be required to make this alternative economical. Vendor reprocessing makes sense as vendors know the constituents present and the best techniques for processing the wastes back to a usable form. Even if only 20 percent were to be reprocessed in this fashion, a considerable pollution prevention benefit would occur. Once wastes are processed, vendors could "bleed-in" the processed wastes into virgin products or sell the reprocessed product, making a "B" grade of the material. The generator may still be able to use the lower grade waste, or it may be sold to other platers with more tolerant processes. In many respects this reprocessing parallels organic solvent recovery firms, which successfully market lower-grade solvents in this fashion.

11.10. Energy Saving. Energy usage and associated costs represent a significant portion of overall pollution prevention costs. As energy generation and distribution creates pollution itself, energy conservation in the metal finishing process and associated waste treatment should be encouraged. In addition, as a direct result of the Clean Air Act Amendments (CAAA) of 1990, energy costs will continue to rise.

Any change to a plating process which uses less current to provide the same level of metal deposition will assist in the overall pollution prevention effort. Off-line soluble metal generators are presently being used to maintain the metal concentrations of plating baths, reducing electricity usage in the cathode bars. From a combustion standpoint, cold black oxide processes can replace hot black oxide tanks in certain applications. Naturally, energy cost savings (natural gas or propane with this black oxide example), must be weighed against other production concerns (such as possible decreased processing time) before implementing processing changes.

12. Limitations and Barriers.

Major barriers and limitations to pollution prevention within the metal finishing industry are listed below.

12.1. Regulatory Issues

12.1.1 Permits. The current regulatory system is not conducive to many potential pollution prevention opportunities. For example, chemical vendor reprocessing of spent products from plating facilities may not be possible without first obtaining a full hazardous waste Treatment, Storage, or Disposal (TSD) permit under the applicable state or federal RCRA regulations. In addition, the permitting process as a whole can be lengthy and costly, particularly for a TSD permit. Years and tens of thousands of dollars may be invested without guaranteed results. Facilities have been constructed but not operated in a timely fashion (or not at all) due to waiting for the necessary permits.

Similarly, acceptance of metal finishing wastes by secondary smelting facilities for eventual return to the primary metals industry is difficult under TSD and hazardous waste regulations. Because EPA considers wastes derived from listed hazardous wastes (such as the typical metal finishing waste treatment sludge listed as F006) as being hazardous, wastes from the secondary smelting operation are also classified as hazardous. As a result, secondary smelters are in large part unwilling or unable to proceed through the long, costly, and intensive maze of hazardous waste requirements.

Many companies simply consider the permitting costs too high, the required length of time too long, the outcome too uncertain, and the public outcry too intimidating to even consider going forward with facility permitting. Unfortunately, good ideas may become abandoned too early.

12.1.2 Conflicting Regulatory Pressures. Ironically, companies can perform a wide variety of pollution prevention techniques and "close the loop" on their plating process only to then be caught in other regulatory issues. One well publicized case involves a Massachusetts company that incorporated a plating line closed loop and eliminated their NPDES discharge [26]. Close looping their process and attaining zero wastewater discharge caused significant concern with other state and federal regulatory agencies. Since their discharges were removed under the Federal Clean Water Act (CWA), their CWA exemption was also withdrawn. As a result, regulatory agencies were requiring this facility to obtain a permit to operate as a TSD facility under the federal and state RCRA regulations. Years later, this company is used as a benchmark success story by one part of the regulatory community while other portions of the regulatory community struggle to determine which regulatory requirements should apply.

The problems of permitting zero wastewater discharge facilities have not yet been settled at the federal level. By law, state standards can be, and many are, more stringent than the federal standards. Even if EPA states a policy or issues regulations on zero wastewater discharges, many states may still require full TSD permits or TSD recycling permits for zero wastewater discharge facilities, especially for facilities using evaporators or other waste concentration methods prior to vendor disposal. Because waste segregation and concentration is required for many pollution prevention opportunities, regulatory conflicts may deter some facilities from changing operations. Given these potential regulatory problems, many facilities will not have the incentive to reach zero wastewater discharge.

12.1.3. Lack of Consistency Between States. State definitions of hazardous wastes and activities requiring permits can differ greatly. Some states may require RCRA permits for certain pollution prevention or recycling processes while others would not. In some cases, waste can be shipped from a generator state where it might be considered hazardous to a state where it is considered nonhazardous. This interstate difference in waste classification (not the actual waste) can limit off-site vendor treatment or disposal options, limit the technologies allowed within the facility due to permitting issues, and change the total economics and incentives of a proposed process change.

12.1.4. Regulatory Differences. Unlike the U.S., many countries lack strong environmental laws and consistent enforcement of these laws. As a result, this non-level playing field provides competitive advantages to metal finishers in these countries. Thus, facilities within these other countries can produce products of equal quality without the regulatory pressures and waste disposal costs experienced by U.S. finishers and may gain an economic edge. As further evidence of these discrepancies, under the CAAA, finishing operations within southern California have very different requirements under the ozone protection section of the law compared to operations within the rural south.

12.2. Expected Cost of Doing Business. Because most companies have already initiated pollution prevention activities, waste reduction rates of 50 percent from the 1980s are common. As a result, total costs for environmental compliance within a metal finishing facility, while still high, are typically much lower today than a few years ago, and these costs are routinely incorporated in company operating budgets. In addition, the significant progress in the past five years and the changing manufacturing environments, the overall economic emphasis for large-scale and immediate pollution prevention costs may not be as strong. Thus, without a concerted corporate policy and goal for pollution prevention (more likely in larger companies), or immediate compliance issues, continued progress in waste reduction may slow. Instead, companies may focus on the issues of scrap, competition, quality, and global marketplace survival.

12.3. Government Specifications. Because changing government specifications is typically a long and expensive process, many metal finishers maintain the status quo rather than implementing innovation or creative engineering.

12.4. Uncertainty of New Products. Suppliers of new products or coatings will naturally attempt to capitalize on their investments as quickly as possible. Sometimes products have been placed into full-scale production before all the "bugs have been worked out." As a result, many larger industries shy away from new products, including those with significant pollution prevention possibilities, until a proven track record has been established. Generally, companies are hesitant to buy expensive equipment with serial 0001, especially if the costs of proven technologies are comparable to innovative ones.

13. Resolution of Barriers and Limitations

13.1. Encourage Trailblazing and Partnerships. Many new technologies or plating products are successfully implemented at small companies with discrete markets and products. These trailblazers may not have the same need to drive products or technologies to market as would larger companies or those who have larger market needs [27].

Innovative ideas or combinations of ideas should be encouraged through partnerships between large and small metal finishers. Joint efforts between component plating suppliers may be able to address specific problems shared by both parties. The larger company may be able to dedicate resources of time, personnel, and computers, while the smaller entity such as a large job shop or a research and development firm may be able to supply the testing operation.

Many advances in chemical substitutions have resulted from cooperation between industry and chemical suppliers. To encourage cooperation, supplier/user partnerships should be established with long-term arrangement that will protect the interests of both parties.

These partnerships may be established by trade groups or perhaps government agencies and encouraged by financial incentives, public recognition, or other positive means. For example, the electric utilities recently sponsored a "contest" to develop a CFC-free refrigerator and used 25 percent less electricity compared to equivalent CFC-containing models. Whirlpool, by making relatively small but significant changes, met the challenge with a new product. Marketing for this new product will likely include mention of the contest specifics, and their successful innovation. Perhaps this type of competition can be sponsored by a trade group with common plating practices.

13.2. Facilitate Permitting. Metal finishing is a highly regulated industry, with permit requirements at the local, state, and federal levels, and permits for air, water, TSD, etc. are often lengthy and costly to obtain. These cost and time "penalties" often deter technology development, as market directions may significantly change before the permitting process is complete. No company, regardless of size, can afford to risk their investment on markets that may disappear. Although some states have begun to recognize barriers to economic development, more states need to expedite the permitting process.

13.3. End Regulatory Schizophrenia. The current regulatory framework focuses on individual discharge media air, water, land rather than facilities as total entities. Perhaps regulations should be changed to view overall satisfactory pollution prevention as regulatory compliance. This global facility approach would help avoid the problems detailed in Section 12.1.2. of a facility that employed positive pollution prevention techniques in the metal finishing discharge area, only to get dragged into regulatory conflicts with other agencies.

As a step forward, EPA has recently started to think in terms multi-media permitting, a positive step.

13.4. Pollution Prevention Grants. Companies are hesitant to place significant amounts of capital into unknown or unproven technologies. Various states understand this concern and have grant money available to help innovative pollution prevention projects. Obviously, expanding the grant programs, increasing their visibility, and minimizing the necessary applications would encourage participation of more companies with innovative technologies and products, especially smaller firms lacking financial and human resources.

13.5. Financial Incentives. Financial incentives such as tax breaks, low interest loans, or other financial incentives can be used to encourage pollution prevention technology development. Thus, incentives might also include "rewards" such as waiving permit fees for facilities that eliminate processes for attempting significant pollution prevention.

13.6. Full Accountability for the Costs of Pollution. A key driving force in pollution prevention programs was the significant and rising cost of pollution and its associated liability. With many of the easily implemented changes already complete, the same financial emphasis and urgency no longer exists. For example, disposal costs of many wastes have stabilized and even decreased in certain areas, due to decreasing waste volumes. Renewed emphasis on pollution prevention technologies will likely occur when the next round of significant disposal costs increases.

In many ways, regulatory programs, such as the CAAA, are increasing the costs of pollution. For example, the CAAA will require permit fees for air pollution, with the fees in each state covering the costs to run the program. Costs for water and sewer fees are increasing significantly in many areas, especially the west. The costs of water sampling (regular and biomonitoring), and permit fees are playing an increasing role in pollution prevention project planning.

As the prices and volumes of wastes have generally decreased, full cost accounting of pollution generation becomes ever more important. In the past, companies could justify projects on easily identified costs; now these costs may need to include more of the background (present and significant) costs in order to truly cost a project.

Thus, metal finishers must include increased reporting requirements and education of EPA or states in the true costs of their pollution prevention. As the metal finishing industry increases its understanding, innovative ideas and vendor/facility partnerships should become much more common.

14. References

[1] Robert K. Guffice, Handbook of Hard Chromium Plating, Gardener Publications, 1986.

[2] Richard Crain, MSFA News Digest, Finishers' Management, November/December 1993.

[3] Products Finishing Directory - 1992, October 1991, Vol. 56, No. 1-A

[4] Metal Finishing Guidebook and Directory Issue - '92, Vol. 90, No. 1A.

[5] Electroplating Engineering Handbook, 4th ed., Lawrence J. Durney, Editor, Van Nostrand Reinhold, New York (1984).

[6] John F. Hanlon, Users Guide to Vacuum Technology - 2nd Ed., Wiley Interscience Publication, 1989.

[7] Hazardous Waste Reduction in the Metal Finishing Industry, PRC Environmental Management Inc., Noyes Data Corporation, 1989, Pollution Technology Review #176.

[8] George C. Cushnie, Electroplating Wastewater Pollution Control Technology, Pollution Technology Review #115, 1985.

[9] USEPA, 1992, Guide to Pollution Prevention - The Metal Finishing Industry, EPA/625/R-92/011.

[10] USEPA, 1982b, Environmental Pollution Control Alternatives: Sludge Handling, Dewatering, and Disposal Alternatives for the Metal Finishing Industry, EPA625/5-82/018.

[11] AESF (American Electroplaters and Surface Finishers), 1991, Workshop II; RCRA/SARA Regulatory Update, Waste Minimization Handouts, AESF Week '91.

[12] Stephen D. Couture, 1984, Source Reduction in the Printed Circuit Industry, Proceedings - The Second Annual Hazardous Materials Management Conference, Philadelphia, Pennsylvania, June 5-7, 1984.

[13] Rolf Kraus, 1988, Shipley Company, Inc., personal communication with Thomas P. Adkisson, PRC Environmental Management, Inc. (April 4, 1988).

[14] Mike Foggia, 1987, Shipley Company, Inc., personal communication with Thomas P. Adkisson, PRC Environmental Management, Inc. (January 21, 1987).

[15] Metal Finishing, 1989, Guidebook and Directory, Vol. 87, No. 1A.

[16] USEPA, 1982a, Control and Treatment Technology for the Metal Finishing Industry-In-Plant Changes, EPAX 8606-0089.

[17] Phil Stone, 1987, Shipley Co., Inc., personal communication with Thomas P. Adkisson, PRC Environmental Management, Inc. (Feb. 24, 1987).

[18] Michael R. Watson, 1973, Pollution Control in Metal Finishing, Noyes Data Corporation, Park Ridge, New Jersey.

[19] A. R. Gavaskar, R. F. Olfenbuttel, J. A. Jones, and T. C. Fox, 1992, Automated Aqueous Rotary Washer for the Metal Finishing Industry, USEPA (in press).

[20] Monica Campbell and William Glenn, 1982, Profit from Pollution Prevention - A Guide to Industrial Waste Reduction and Recycling, Pollution Probe Foundation, Toronto, Ontario.

[21] USEPA, 1985, Environmental Pollution Control Alternatives: Reducing Water Pollution Control Costs in the Electroplating Industry, September 1987, EPA 625/5-85/016.

[22] George D. Mitchel, 1984, A Unique Method for the Removal and Recovery of Heavy Metals from the Rinse Waters in the Metal Plating and Electronic Interconnection Industries, Proceedings - Massachusetts Hazardous Waste Source Reduction, Clinton, Massachusetts.

[23] Frank Altmeyer, Advice and Counsel - Contamination Prevention: Both Sides, Journal of Plating and Surface Finishing, Vol. 79, No. 8, pp. 23-25.

[24] Robert B. Pojasek, Finding Safer Materials May not be as Easy as you Think, Pollution Prevention Preview, Winter 1993-94, Vol. 4, No. 1, pp. 119-122.

[25] Kevin E. Warheit, Zero Discharge - Fact or Fiction, A Waste Minimization Case Study, Presented at the American Electroplating and Surface Finishing Environmental Symposium, 1992.

[26]   The Robbins Company: Wastewater Treatment and Recovery System, A Case Study, Massachusetts Department of Environmental Management, Office of Technology Assessment, Boston, MA.

[27]   Marty Borusco, Finisher's Think Tank, Journal of Plating and Surface Finishing, January 1994, Vol. 81, No. 1, p. 34.

About the Author

Kevin P. Vidmar is the Division Manager of Environmental Affairs and Plant Services for Stanley Fastening Systems, a division of The Stanley Works. He has published and presented on waste minimization at numerous conferences. At both the 1991 and 1992 annual environmental symposiums of the American Electroplating and Surface Finishing society, Mr. Vidmar received the award for Presentation of the Year. Mr. Vidmar received a B.A. in Zoology from Miami University (Ohio), and an M.S. in Environmental Engineering from Vanderbilt University. In his present capacity, he is involved with the environmental activities for 10 manufacturing locations, in addition to facilities maintenance, machine installation, and machine repair.

Pollution Prevention in the Pulp and Paper Industry
David H. Critchfield
Manager of Regulatory Affairs
International Paper
Jay, Maine 04239
(207) 897-1605

Government regulation and limits on industrial emissions form the backbone of the environmental control system in the Western Hemisphere. This likely will continue to be true for at least five more years. However, several key changes have increased the likelihood that pollution prevention will play an important role in shaping the U.S. paper industry's competitiveness in the world. Companies are making available to employees, customers, and the public more information about the amounts of chemicals and toxic substances used in manufacturing their products. This fact inspires efforts to increase the efficiency of manufacturing processes, thereby reducing waste prior to disposal. Innovative government and industry programs are redirecting R&D support towards technologies which promise better energy and raw material-to-product consumption ratios. Finally, smart manufacturers realize it is a very short step indeed from environmentally friendly products which the consumer is demanding to environmentally friendly processes.

Key Words

Air emissions; bleaching; chlorine; closed-loop recycling; Cluster Rule; dioxin; input substitution; mini-mill; pollution prevention; process modification; pulp and paper industry; recovered paper; wastewater.

1. Introduction

This paper highlights some of the key opportunities and needs that small- to medium-sized businesses targeting the U.S. pulp and paper industry with pollution prevention technologies can fill.

Historically, pollution control technologies (end-of-pipe recycling) have been to engineering graduates what Le Morte d'Arthur and Beowulf are to students of English literature: worthy and revered, but a little dull. No longer. Pollution prevention has shifted the focus of environmental engineers and regulators from traditional waste treatment to methodical examination of the efficiency of the manufacturing process itself. Eliminate waste from the manufacturing process and thereby avoid or reduce significantly the need to treat waste using traditional, energy-intensive end-of-pipe methods.

This approach seems so logical from a process design perspective that one cannot help but wonder why it has taken almost 25 years for industry, regulators, and the environmental movement to actively champion pollution prevention. The answer to that question will perhaps become clearer when we examine some of the opportunities for small business to provide solutions to the pulp and paper industry.

A second question must also be answered if we hope to be able to rank and select the best candidates from the broad array of opportunities facing U.S. pulp and paper manufacturers. What are the key factors which will affect U.S. manufacturers as they enter the 21st century, and how will they maintain their hard-won position of world leadership in technical innovation, productivity, and product quality? These questions cannot be answered without some understanding of the breadth and depth of the industry, its products, and the way it is currently regulated by EPA and the states.

In 1992, the U.S. pulp and paper industry shipped $131 billion worth of product to customers worldwide. Overall, approximately 4.5 percent of all manufacturing shipments originated from this country. Of that figure, $10.3 billion was exported. This level of economic activity was carried on by 544 mills at latest count, operating in 42 states.

The industry produced 82.9 million tons of paper and paperboard in 1992, the most recent year for which data is available. Industry capacity that year was 88.8 million tons, which is approximately 30 percent of total world capacity. In the U.S., 57 percent of the productive capacity was less than 10 years old in 1991. The amount of waste paper recovered for recycling in 1992 was 26.2 million tons. This amounts to 31 percent of production and 38.1 percent of all paper consumed in the U.S.

Industry energy efficiency has improved steadily since the late 70s. In 1991, the pulp and paper industry consumed 2.55 quadrillion Btus, 56 percent of which was self-generated. This constituted 3.1 percent of total U.S. energy consumption.

2. Discussion

2.1 Framework for Analysis. At a 1993 *Mill of the Future* workshop, Dr. Delmar Raymond, Director of Energy Science & Technology for Weyerhaeuser, pointed to five key factors he thought would affect the success of the U.S. pulp and paper industry [1]:

> Diversion of capital from production-related investments to environmental control systems
> Shift in mill steam/power balance to increased dependency on purchased electric power
> Increased global competition

Increased use of joint ventures with utilities and others for electrical production

Increased use of R&D partnerships to make technical innovations commercially viable.

Arguably, Dr. Raymond is exactly right. This author could suggest only one additional factor that perhaps would make the list more complete: the increasing customer demand for products manufactured in an environmentally responsible manner. This phenomenon has, and will continue to, influence the way manufacturers make their products and how they allocate scarce capital to cover other competing needs such as productivity, quality, and environmental compliance.

Each of these issues by themselves would justify a thorough analysis well beyond the scope of this paper. Yet, taken as a whole, they point out the need for some kind of analytical framework, to support understanding of which opportunities are most promising and which needs are most pressing. In this way, businesses can perhaps develop a strategy for marketing pollution prevention technologies to the pulp and paper industry.

Only a short time ago, otherwise reasonable people might nearly come to blows debating which manufacturing activities actually constituted pollution "prevention," and which were really pollution "reduction." One might be inclined to laugh. Nevertheless, serious people, regulators, and private sector professionals have put much energy into this debate, frankly to little avail. While this paper does not purport to answer this question, it necessarily puts forth a list of manufacturing practice categories which should be considered pollution prevention.

Input substitution (e.g., secondary fiber substitution)
Product reformulation (e.g., eliminate chemical usage requiring TRI reporting)
Process modification (e.g., change pulping or bleaching methods)
Improved house-keeping, and
Closed-loop recycling

In setting priorities for R&D and investment, corporations employ a variety of models. Small businesses might try similar decision-making processes in deciding how to develop and market their ideas to the pulp and paper industry. For example, a model which uses the following criteria (not necessarily ranked in order of importance) might help a mill process engineering department prioritize limited capital spending resources.

Health and environmental risk
Technology risk
Capital requirements
Company and regulatory resource requirements
Customer expectations

2.2. Pollution Prevention - Preface on Industry Challenges. There are opportunities for small- to medium-sized businesses to participate in the pollution prevention activities within the pulp and paper industry. These opportunities are moderate, at best, for companies planning to break into the markets supplying the paper industry for several key reasons as discussed below.

EPA's Cluster Rule. For at least the next four to six years, the paper industry will be analyzing and complying with the new EPA regulatory proposal known as the "Cluster Rule," so named because it groups or clusters new regulatory initiatives for both air and wastewater discharges into a single regulatory package. Just published as a proposal in December 1993, this rule will not only mandate tougher effluent and air emission limits, but will also specify manufacturing technologies to be substituted for current processes such as pulp bleaching [2]. Although these requirements are not expected to become effective until 1995 at the soonest, just being proposed will measurably impact planned corporate capital spending and allocation of funds. The EPA rule itself was published and justified in the name of pollution prevention, but it will doubtless constrain the ability of industry designers to target more promising ways to cut pollution in meaningful ways.

Recycling Capacity Continues to Increase. The dramatic increase in consumption of recovered paper since 1988 through new capacity also has diverted available capital from other possible uses. Certainly, recovery of waste paper has had a beneficial impact on life expectancy of municipal landfills across the country. Studies have yielded conflicting results as to whether the life-cycle costs in real dollars are better with recycling versus other management approaches (e.g., composting or incineration). That debate will no doubt continue, but the impact on the pulp and paper industry, in terms of capital allocation, will likely have both positive and negative implications for the innovation and world competitiveness of the U.S. paper industry.

Barriers to Entry - Few Industry Suppliers. Although the pulp and paper industry represents a significant portion of total U.S. manufacturing (i.e., 4.5 percent of all 1992 shipments), relatively few domestic and international suppliers provide design, engineering, and fabrication services. This gap presents barriers to entry for small- to mid-sized firms not already selling goods and services to this industry. It is not an insurmountable hurdle, but means that small firms must target their customers carefully and not spread themselves too thin by trying to reach all the major customers at the same time.

2.3. Opportunities and Needs in the Industry. Despite these obstacles, clearly there are opportunities for small companies to help the paper industry find and capitalize on pollution prevention initiatives. Pollution prevention efforts have demonstrated savings in product and waste treatment costs and are fast becoming one of the more important ways a company can give tangible proof to claims of environmental "sensitivity." Of the five pollution prevention concepts advocated by the EPA (discussed in Sec. 1), four directly impact manufacturing areas as opposed to waste treatment operations. Firms that originally provided only services or products which focused solely on waste management should view this as a chance to access the rest of the mill, to reach different customers under the same roof.

One way to systematically address the opportunities is to look at each category. Because there are so many different ways to eliminate waste, we will turn back to the concept suggested above, focusing on the areas that promise the highest "return" in terms of five criteria: health and environmental risk, technology risk, capital, resources, and customer expectations.

Input Substitution. Substitution of recovered paper for virgin fiber in paper making has increased from approximately 22 percent in 1988 to more than 40 percent today. This change has had an enormous impact on this industry and, although it would score relatively low in the area of health or environmental risk, inarguably it would score quite high on the remaining four criteria, especially customer expectations. Some paper companies have been using recovered paper for many decades, but for others the market is new. Figure 1 illustrates a typical schematic for an integrated pulp & paper manufacturing facility. Figure 2 shows schematically where recovered paper would be introduced. For a typical integrated mill with batch or continuous digesters, the cost of shipping and processing recovered fiber still exceeds the cost of own-make pulp, since most integrated mills are located far from sources of waste paper. Nevertheless, firms supplying goods and services to the formerly small niche recycling companies should now find their potential markets expanding, and the demand for innovation and quality engineering growing as more mills begin recycling.

Perhaps the most dramatic change brought on by recycling is the emergence of the mini-mill manufacturing concept [3,4,5]. Recalling a similar phenomenon in the steel industry 15 years ago, mini-mills in the paper industry are a natural outgrowth of the trend to want the use of more recycled fiber in the paper industry. Companies choosing to provide products with recycled content realized several advantages in scaling down and relocating their capacity closer to urban areas, their principal source of raw material supply. Lower transportation costs, fewer environmental licensing requirements, and reduced environmental impacts through design and scale effects have been demonstrated by recent mini-mill projects [6].

The industry forecasts around 50 percent of all paper manufactured by 1998 or approximately 50 million tons per year will be recovered for recycling at new or rebuilt existing mills. Further innovations will be demanded in key production areas to improve the economics of the mini-mills built to convert some of that additional tonnage. These will include:

    Water conservation and reuse systems
    Process control systems
    Improved energy and steam systems.

Product Reformulation. Because of public and regulatory pressures brought on by recent programs such as EPA's Toxic Release Inventory (TRI), for several years chemical suppliers to the paper industry have been making product changes and reformulations. These measures are intended to help customers by reducing the quantity of chemicals characterized as toxic in the manufacturing process, and thus to reduce the amount that must be reported to agencies and that may potentially reach the environment if improperly handled. There still are opportunities for reformulation of some products, but perhaps the greatest potential for environmental improvement rests with conservative use of existing formulations and elimination of product waste and spillage.

From a health and environmental risk and customer expectation perspective, improvements here would be beneficial. Still, the greatest gains for most mills are associated with process changes, as discussed below. Methanol and chloroform are both TRI reportable chemicals, and their by-product status will be altered only by treating foul condensates from the pulp and bleach plant portion of the integrated mill. Compared to other aspects of the manufacturing process, this area does not hold as much promise for market development by the small to mid-sized firm at this time.

Process Modification. Without question, the most discussed and documented subject in the paper-making industry is the use of chlorine in pulp bleaching. The discovery in 1985 of the chemical congeners of dioxin in pulp mill effluents sparked the intensive research and development efforts by the paper industry and government organizations which continue today. It also kicked off a competitive race within the industry to develop bleaching technologies that produce the fewest possible chemical by-products that might adversely impact the environment. Now, almost a decade after dioxin was first linked to paper making and other processes such as incineration, the strong debate continues over which bleaching technologies promise the best outcome in terms of environmental quality impacts [7]. This issue highlights some of the questions EPA is now beginning to ask about science-based decision-making and environmental protection policy. The balancing of environmental protection goals with those for economic development, against a backdrop of scientific uncertainty, will prove to be a challenging social issue [8,9].

The EPA joined the debate this past December by taking the risky and controversial step of specifying what process technologies the industry ought to adopt in the Cluster Rule, which established new air and water discharge limits [2]. Earlier in the year, EPA published an extensive technology review entitled, Pollution Prevention Technologies for the Bleached Kraft Segment of the U.S. Pulp and Paper Industry [10]. A companion to the same document was published as a handbook last June [11]. Both publications were devoted almost exclusively to the economics and environmental impacts of new pulping and bleaching technologies.

Not surprisingly, this intensive period of research and regulation and the accompanying media coverage have caused considerable confusion and disagreement over what should be done. The industry, its customers, the public, and regulators clearly do not share a common vision of the measures that will be necessary ensure that paper products continue to be provided to the marketplace using the most environmentally friendly processes possible. Obviously, processes that cause the least amount of the pollution before treatment are preferable. Unfortunately, there is not yet consensus on which processes cause the least amount of pollution on a life-cycle basis.

Is this then a promising area for small businesses to explore? Perhaps. The stakes are quite high, and the active players in pulping and bleaching technologies tend to cut a wide path through the competitive jungle corporate R&D units, well-funded university research facilities, industry partnerships. Moving ideas from the laboratory to full scale requires considerable amounts of risk capital. This forces smaller firms with promising ideas to search for partnerships.

The best prospects will likely fall to those firms that currently provide environmental services in the form of compliance-driven waste treatment systems. Engineering services firms which spend considerable time with mill clients solving compliance problems are well positioned to identify process-related opportunities which have a reasonable payback in terms of waste elimination, but are too small to attract the attention of corporate researchers. Competitive advantage will accrue to those pulp and paper firms that search out every possible waste reduction measure before turning their attention to designing waste treatment systems to comply with the new Cluster Rule. Many such waste-reduction measures will have relatively low technology risk and capital requirements, but how many of them are implemented will depend on the limited resources of client mills.

Improved Housekeeping. Integrated pulp and paper mills, and even the newer mini-mills, are large and complex manufacturing facilities. All modern mills run as continuous processes. Process upsets, therefore, have the potential to trigger product losses and waste throughout the facility, because the rest of the operation cannot shut down while the root cause of the problem is being identified and rectified. Improved housekeeping maintaining a clean and orderly operation greatly reduces the chances that contamination will cause upsets. This is not a glamorous aspect of pollution prevention, compared with process modifications for example. It falls more in the area of "attention to detail," which does not lend itself as much to the services of small firms as it does the discipline and pride of mill employees.

Employee performance and productivity is at the core of any successful organization. Recognizing this fact has prompted many companies to shift their management philosophy away from an autocratic, top-down driven approach toward self-directed, team-oriented, high-performance efforts.

Closed-loop Recycling. At the bottom of most government pollution prevention lists, closed-loop recycling often constitutes waste treatment or recovery and reuse after process losses already have occurred. The importance of this method should never be discounted. It is both an important part of the pollution prevention equation and an opportunity for small firms to gain access to the pulp and paper market. In the past few years, and for several years to come before the Cluster Rule goes into effect, the inspiration for process changes that prevent pollution altogether has originated with a waste treatment problem. Young engineers

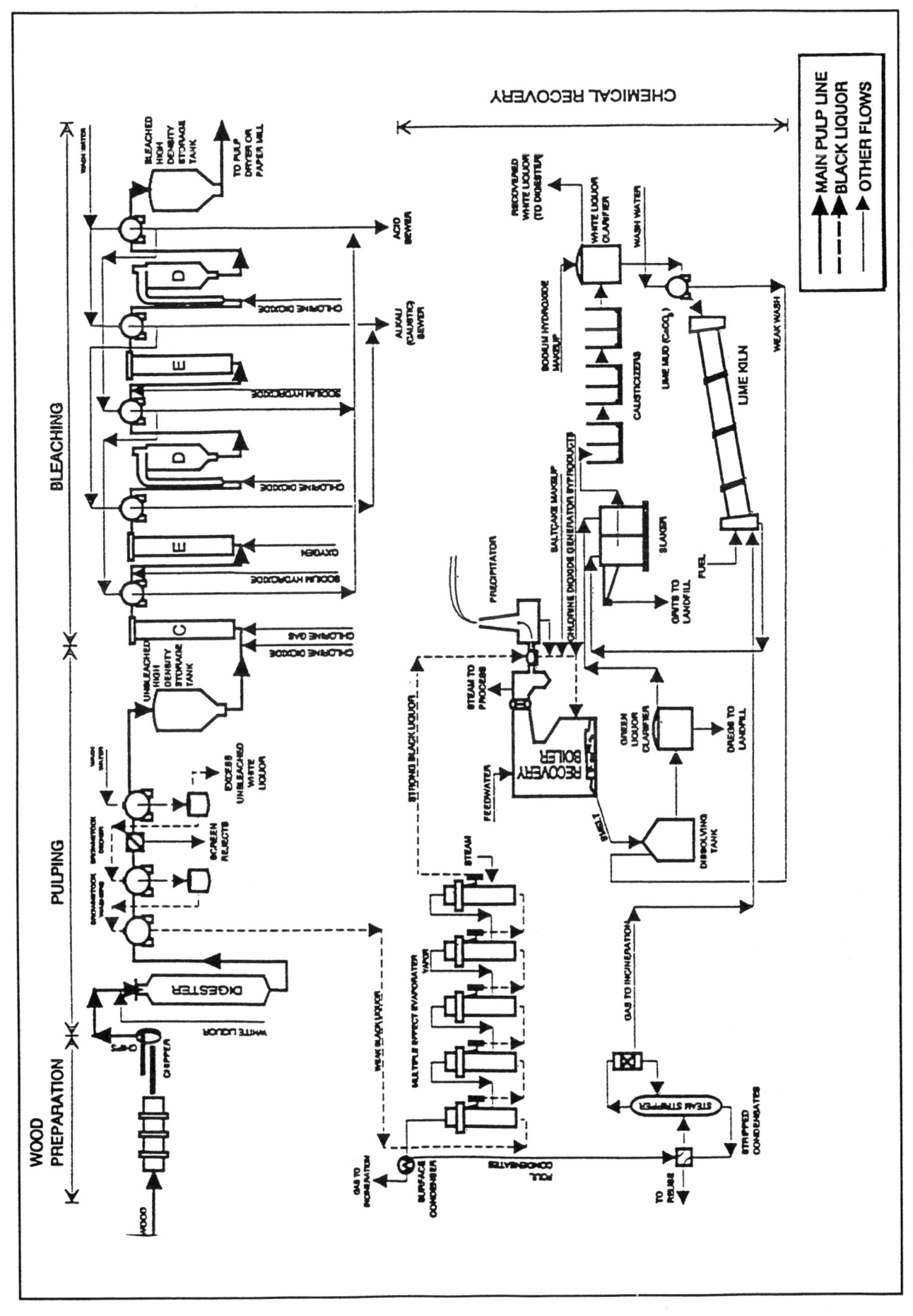

Figure 1. Flowchart for Conventional Integrated Bleached Kraft Mill (Mid- to Late- 1980s)

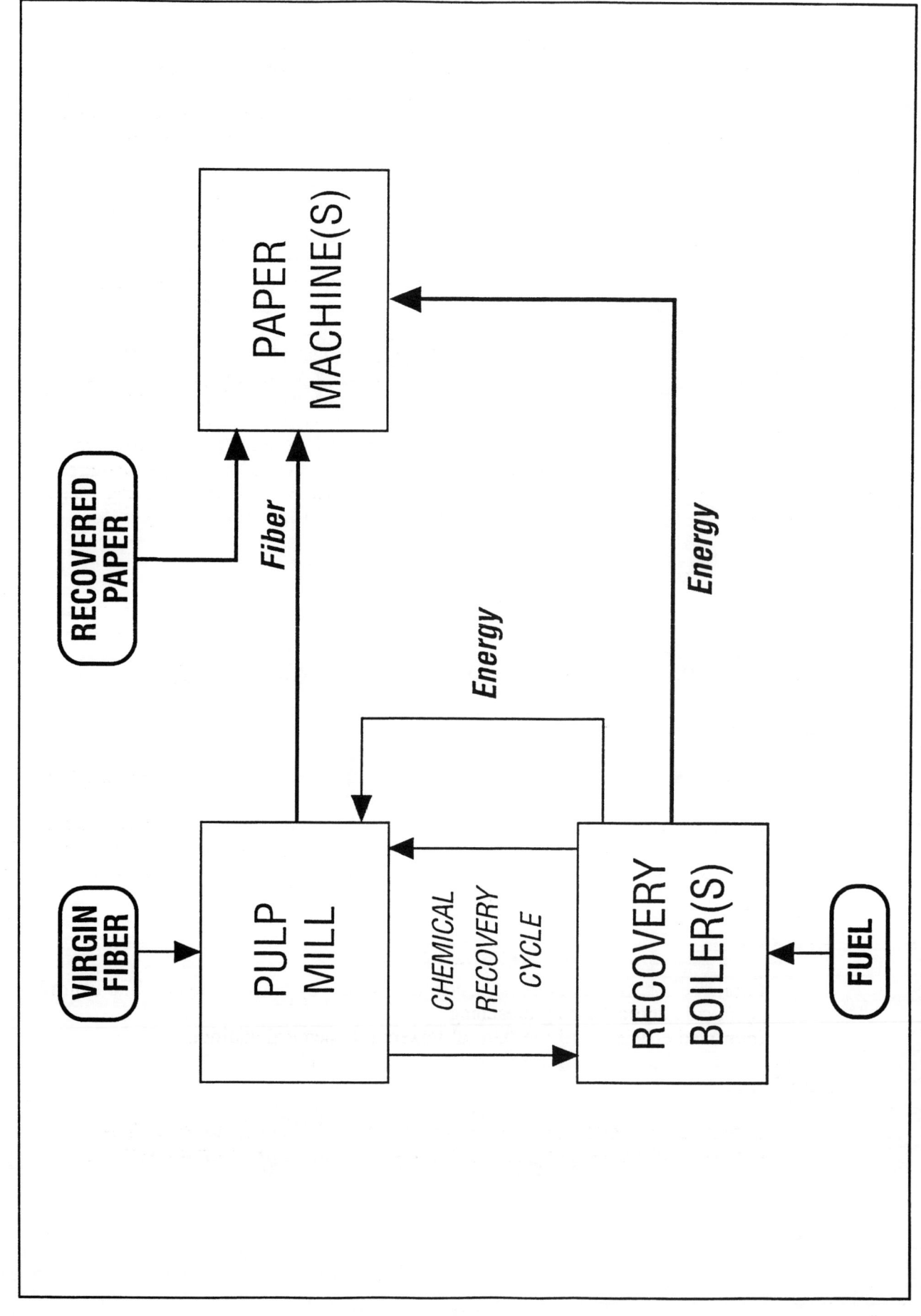

Figure 2. Block Flow Diagram Illustrating Introduction of Recovered Paper into Manufacturing Process

and creative engineering firms did not allow their thinking to be cramped by a narrow definition of the problem. They have challenged their clients to make an important linkage which is critical to any successful pollution prevention program: one cannot manage what one doesn't measure. Better measurements, and more extensive documentation of process losses, have led to better understanding of the cost and energy tradeoffs between treatment and prevention.

Again, the opportunities fall most to those firms that have forged reputations in the design and installation of waste treatment and air emission control systems. These firms have the experience and skilled engineering talent to redefine a waste treatment problem as a waste minimization problem. In that regard, the EPA Cluster Rule represents an opportunity as well as an obstacle for this industry. It is an opportunity in that mills hopefully can take advantage of the lead time to analyze different combinations of process and waste treatment system changes, searching for the best economic and environmental solution.

It is apparent from the discussion above that, in the coming years, the U.S. paper industry will have to walk a very narrow tightrope to balance reinvestment for quality and capacity against EPA's Cluster Rule and increased public expectation that industry will continue to reduce its environmental profile. Hence, the recent enthusiasm for *pollution prevention*. Obviously, the areas with the greatest promise for investment and research are those where all three objectives can be met. For small- to medium-sized businesses that are persistent and innovative, the following manufacturing areas should yield results. Long-range research and development initiatives, such as black liquor gasification, have been omitted.

Wood Yard Operations

> Improved log debarking measures including ring debarking, reduced hydraulic loading in debarking operations, non-liquid lubricants in wood yard operations
> Chip quality improvement measures including extended blade life, thickness controls, reduction of rejects, and the amount of shives in accepts

Pulping and Chemical Recovery

> Spill prevention and control measures, rapid detection and containment of liquor or brownstock losses in the pulp mill
> Improved and new brownstock washing systems, better recovery of liquor solids
> More efficient systems to strip volatile compounds from foul condensates
> More efficient systems to capture and eliminate odorous gases such as total reduced sulfur compounds from pulping operations

Pulp Bleaching

> Improved instrumentation and distributed control systems to monitor and control chemical addition during bleaching
> Improved efficiency of mechanical mixers for chemical addition

Pulp Drying

> Improved dust collection systems for capture of difficult fugitives such as hardwood fiber, and recovered post-consumer fiber in flash-drying systems

Papermaking

> Improved fiber and coating recovery systems for paper mill sewers
> Improved fiber saveall systems for paper machine white water

Wastewater and Solid Waste Management

> More efficient systems to maximize solids content of dewatered sludge with increasing ash content
> Energy-recovery methods using wastewater sludges such as through gasification
> Economically attractive methods to handle relatively uncontaminated process wastewaters, without adding hydraulic loading to costly wastewater treatment systems

3. Conclusion

The U.S. pulp and paper industry is struggling to break free of the recession that has gripped this country and the world for the past four years. Environmentalism has greatly increased its breadth and depth during the same period, manifesting itself in a strong and perhaps permanent demand for environmentally friendly products. Government promises to increase its involvement and control over this industry, in the form of new requirements that extend beyond the degree to which wastes are treated, to address regulating the way paper is made. These three factors put the pulp and paper industry on the threshold of enormous change. When this industry emerges from this period of change and crosses into the next century, the companies that assume leadership roles will likely be those that were most open to ideas and innovations from outside of the industry. The small- to mid-sized firms that help this industry find the successful pollution prevention ideas will also be well positioned to take a leadership role in their respective markets.

4. References

[1] D. R. Raymond, The Mill of AD 2020 - Challenges and Opportunities, Pulp and Paper Mill of the Future Workshop, University of Maine, September 9, 1993.

[2] U.S. Environmental Protection Agency, Proposed Cluster Rule, Federal Register, 58 FR 66078, December 17, 1993.

[3] R. W. Sackellares, Environmental Issues Play Key Role in Planning, Siting Urban Mini-Mills, Pulp & Paper, September 1993.

[4] G. Fales, U.S. Rediscovers the Mini-Mill, World Paper, December 1993.

[5] M. Koepenick, Mini-Mill Relies on Corr Community, PIMA Magazine, January 1994.

[6] R. Hoffman, Sizing and Locating the Containerboard Mill of the Future, OCC Compendium, Pulp & Paper, 1993.

[7] B. Hileman, Concerns Broaden over Chlorine and Chlorinated Hydrocarbons, Chemical & Engineering News, April 19, 1993.

[8] R. Stone, Can Carol Browner Reform EPA?, News & Comment, Science, January 21, 1994.

[9] I. Amato, The Crusade Against Chlorine, News & Comment, Science, July 9, 1993.

[10] U.S. Environmental Protection Agency, Pollution Prevention Technologies for the Bleached Kraft Segment of the U.S. Pulp and Paper Industry, Office of Pollution Prevention and Toxics, Washington, D.C., EPA/600/R-93/110, August 1993.

[11] U.S. Environmental Protection Agency, Handbook on Pollution Opportunities for Bleached Kraft Pulp and Paper Mills, Office of Enforcement, Washington, D.C., EPA/600/R-93/098, June 1993.

About the Author

David Critchfield is the Manager of Environment at International Paper, Androscoggin, Maine. He is responsible for environmental compliance programs at the 1,500-ton per day integrated pulp and paper mill facility.

Pollution Prevention in the Printing Industry
C. Nelson Ho, Ph.D.
Manager, Corporate Environmental and Safety Services
Graphic Arts Technical Foundation
Pittsburgh, Pennsylvania 15213
(412) 621-6941

The printing industry generates three major categories of pollution in large quantities: (1) volatile organic compounds (VOCs) and air toxics emissions, (2) process wastewater, and (3) industrial and hazardous wastes. These types of pollution are generated from the use of various types of photoprocessing chemicals, printing inks, cleaning and wash-up solvents, lubricating oils, and adhesives during the printing operation. This article describes some of the most significant environmental problem areas that printers face as well as the exciting opportunities that the printing industry is witnessing today. Six distinct measures may reduce and even prevent pollution from being generated. These, too, are explored in detail.

Key Words

Artificial intelligence; chained and blown up effect on waste generation; chained and irreversible process; makeready; prepress, press, and postpress operation; printing process; total quality management.

1. Introduction

The printing industry is one of the top ten industries in the U.S. With more than 70,000 operating plants, its facilities outnumber all the other industries [1]. It is mainly composed of six major branches that can be categorized according to the printing process they involve. These processes include: sheetfed offset; heatset web offset; nonheatset web offset; gravure, flexography, and screen printing; smaller sub-branches such as letterpress and thermography; and many different combinations of these printing processes in one plant (e.g., heatset web and sheetfed or sheetfed and flexography.) Figures 1 to 6 illustrate these various printing processes [2].

Each branch of printing is unique in its distinctive prepress, press, or postpress operations. Each process in every branch uses a variety of raw materials and chemicals. In the prepress operation, a variety of photoprocessing films, plates, and chemicals are used. The main concerns in this operation are spent photoprocessing chemicals such as film fixer and developer and plate developer and finisher (process wastewater), which are routinely discharged into sewers by most printers operating in large metropolitan areas [2].

The press operation involves many different grades, shapes, and sizes of paper stock; printing inks, overprint varnishes, and coating materials; straight and blended organic cleaning solvents; types of plate and blanket cleaning and preserving chemicals; and grades of lubricating oils. The main environmental concerns in this operation include stack and fugitive VOC emissions (i.e., air pollutants) generated from the use of inks and cleaning solvents; the wash-up waste (i.e., industrial and hazardous wastes) generated during routine roller, plate, and blanket cleaning; waste and off-spec inks (i.e., industrial and hazardous wastes); huge quantities of printed waste papers (mostly recycled); and the waste oils (mostly used for heat recovery) generated from lubricating of printing presses [2].

In the postpress operation, large quantities of various grades of adhesives are used in the binding of printed products. The main environmental concerns in this operation include VOC emissions from solvent-based adhesives and large quantities of rejected printed products [2]. The largest contributor to pollution in postpress operation is the paper waste generated as well as everything else that is involved in making that waste paper.

Most of these concerns VOC emissions, industrial and hazardous wastes, and wastewater are mainly due to the overuse of raw materials, lack of total quality management, failure to recognize the benefit of pollution prevention to the overall business, and lack of stimulus for initiating changes.

2. Reduction and Prevention of VOC Emissions

2.1. A Semiautomatic, Portable, and One-Size-Fits-Many Blanket/Roller Washing and Cleaning Device. Most printers still wash and clean the ink roller train, blanket, metering roller, plate, cylinder, and screen manually using organic wash-up solvents during job changes, ink color changes, or clearing of minor printing problems. The pressman performs the cleaning process when called for. To do so, he simply dips a piece of cloth rag into an organic solvent, normally a low flash solvent blend (e.g., xylene and isopropyl alcohol). The rag absorbs the solvent and is then carried to the press for cleaning. The bigger the press, the more solvent is used. Routinely, the soaking wet rag drips on the way to the press or is drained on a pile of rags to get rid of the excess solvent before it is carried to the press. In the process, a large amount of the solvent evaporates into the pressroom and eventually leaks out of the building. Thus, it becomes an emission source of VOC, an ozone precursor in the lower atmosphere. This manual cleaning process tends not only to dramatically overuse the wash-up solvent but also wastes a large amount of manpower and machine time. All printers understand this because they know the process is costly in terms of money and resources. Several printing press manufacturers have successfully produced automatic blankets and roller was hers that are permanently mounted on their newly manufactured presses. Older presses have to be retrofitted to permit the use of such a device. This is costly and requires considerable press downtime. Most printers have older presses, some of which are difficult to retrofit. Printers would love to change this manual chore with some sort of light, portable, retractable, and one-size-fits-many device that can be adjusted to fit their printing presses and achieve the same result as a pair of human hands. Unfortunately, no such device is available. Note that printing presses vary in size. Smaller presses are approximately 6" to 18" wide; medium presses are 22" to 48" wide; and large presses range from 50" up. Presses also come in different forms and shapes. If such a device can be developed, it certainly will reduce and prevent large quantities of VOC from being released to the atmosphere by reducing the use of organic wash solvents, especially in the major metropolitan areas since this is where most printers are located. In addition, such a device definitely will save printers substantial sums of money.

2.2. A Universal Wash-up Solvent. Printers must use wash-up solvent to clean the roller train, blanket, plate, screen, cylinder, and other pieces of equipment that come into contact with printing inks composed mainly of hydrocarbon oils, vegetable oils, organic solvents, resins, rosins, pigments, water, and hundreds of different additives. There are millions of ink formulae currently used by ink manufacturers. Each is different. This makes cleaning very difficult. Most printers normally use a blend of alcohol and aliphatic and aromatic solvents (usually 100 percent VOC), that has a fairly low flash point (usually <140°F). The blend cleans the ink residues very quickly, evaporates almost immediately, and leaves no residue. Because of the high evaporation rate, much of the wash-up solvent is lost in the pressroom air before it is effectively used, especially during makeready as well as job and color changes. Consequently, VOC emissions are generated.

Many organic solvent manufacturers have developed various low VOC wash-up solvents for printers. So far, none has been very successful in meeting printers' demands. Many solvents have fairly high flash points (>140°F and still 100 percent VOC), but they leave too much residue after each use. Thus, printers have to use another solvent with a low flash point to clean the residue and, consequently, spend more time cleaning blankets, rollers, and plates. The others have very strong, sometimes offensive, and distinctive odors that can cause pressmen to become ill.

Printers need a solvent that has a fairly high flash point so it will not flash so quickly and disappear into the pressroom during application. Such a solvent can still clean ink residue as effectively as those with low flash solvents do, yet would leave no residue after application. Most critical of all, no further cleanup would be needed if such a solvent was used. If this ideal wash-up solvent could be developed, printers would definitely be able to greatly reduce and even prevent the air pollution when using wash-up solvent for cleaning. In addition, they would be able to greatly reduce the cost of purchasing wash-up solvents.

3. Integrated Waste Minimization and Pollution Prevention

3.1. Total Quality Management. The printing process is different from many other industrial processes and also unique by virtue of its own peculiar characteristics. The most prominent of these characteristics is the "chained, irreversible process" [3]. As indicated in Fig. 7 which uses sheetfed offset lithography as an example, printing is a step-wise process where each step resembles a link in a piece of chain. Each step produces a product that leads and feeds to the next one until the final product is made. The product from each step is not reversible once it is made. If a flaw is found in the product, it almost has to be remade. If a

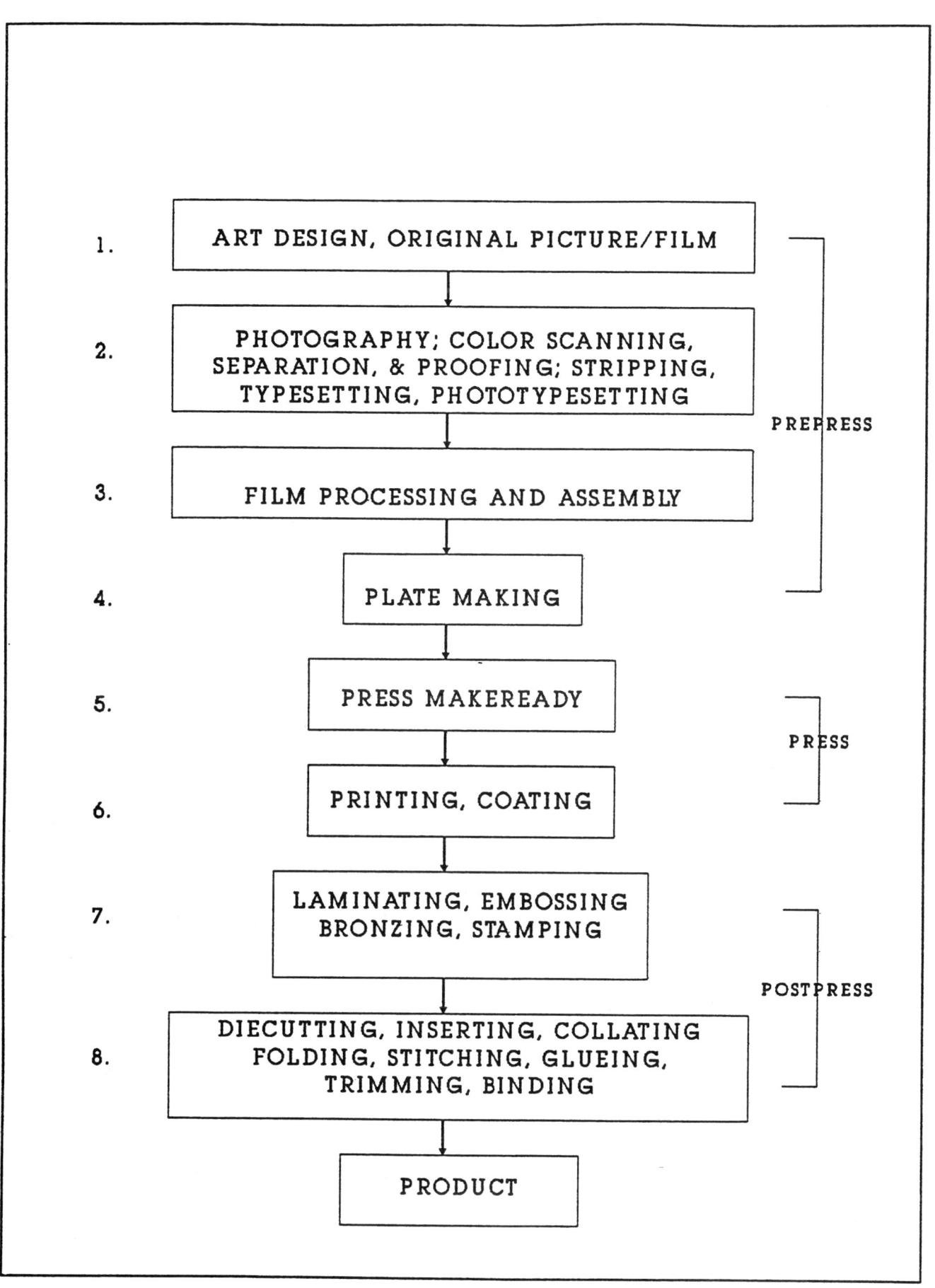

```
                    PREPRESS
1.  ART DESIGN, ORIGINAL PICTURE/FILM
2.  PHOTOGRAPHY; COLOR SCANNING,
    SEPARATION, & PROOFING; STRIPPING,
    TYPESETTING, PHOTOTYPESETTING
3.  FILM PROCESSING AND ASSEMBLY
4.  PLATE MAKING

                    PRESS
5.  PRESS MAKEREADY
6.  PRINTING, COATING

                    POSTPRESS
7.  LAMINATING, EMBOSSING
    BRONZING, STAMPING
8.  DIECUTTING, INSERTING, COLLATING
    FOLDING, STITCHING, GLUEING,
    TRIMMING, BINDING

    PRODUCT
```

Figure 1. Sheetfed Offset Operation Process Flow Diagram

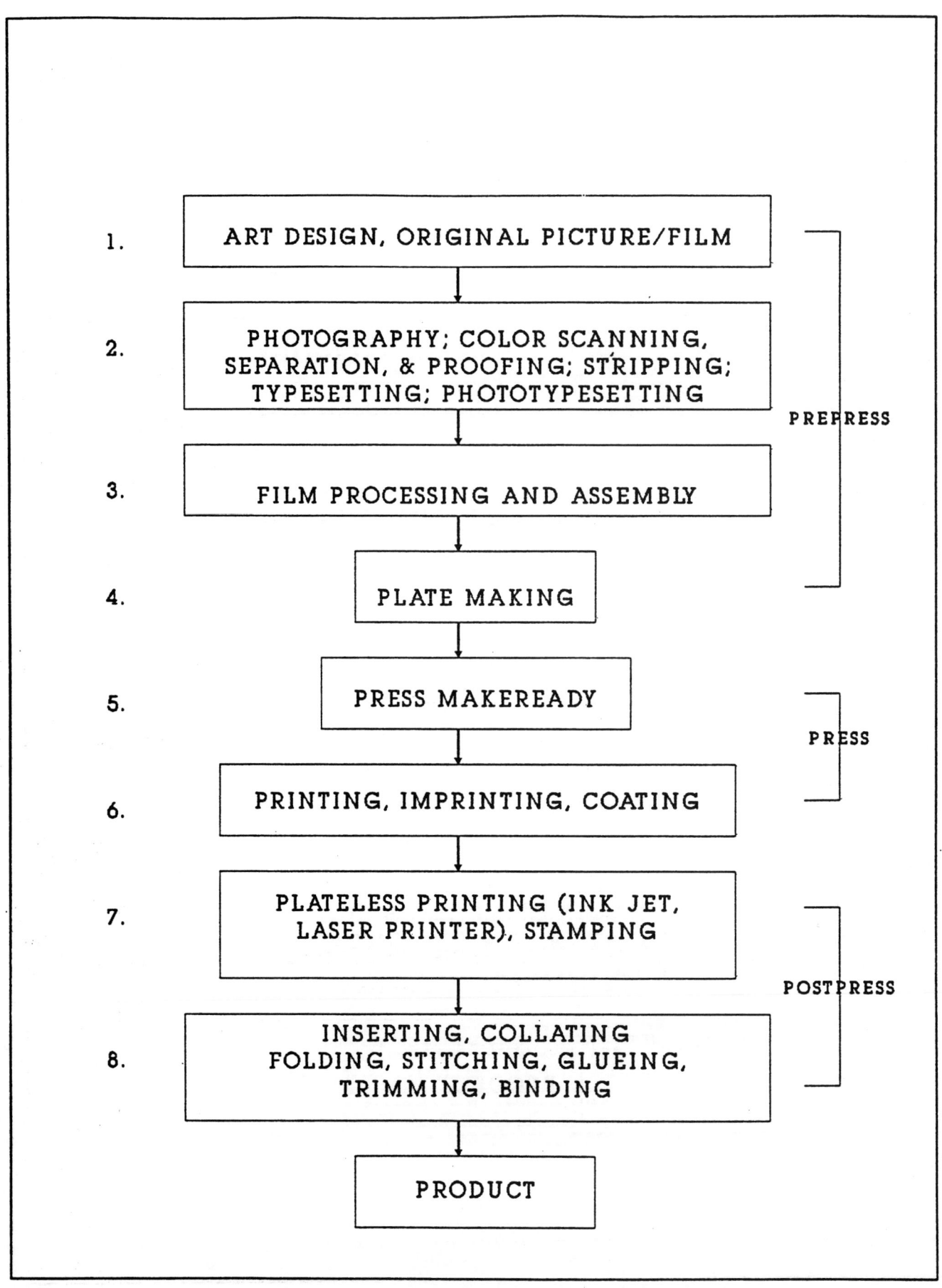

```
                                                    ┐
1.   ART DESIGN, ORIGINAL PICTURE/FILM              │
                                                    │
2.   PHOTOGRAPHY; COLOR SCANNING,                   │
     SEPARATION, & PROOFING; STRIPPING;             ├ PREPRESS
     TYPESETTING; PHOTOTYPESETTING                  │
                                                    │
3.   FILM PROCESSING AND ASSEMBLY                   │
                                                    │
4.   PLATE MAKING                                   ┘

5.   PRESS MAKEREADY                                ┐
                                                    ├ PRESS
6.   PRINTING, IMPRINTING, COATING                  ┘

7.   PLATELESS PRINTING (INK JET,                   ┐
     LASER PRINTER), STAMPING                       │
                                                    ├ POSTPRESS
8.   INSERTING, COLLATING                           │
     FOLDING, STITCHING, GLUEING,                   │
     TRIMMING, BINDING                              ┘

     PRODUCT
```

Figure 2. Heatset Web Offset Operation Process Flow Diagram

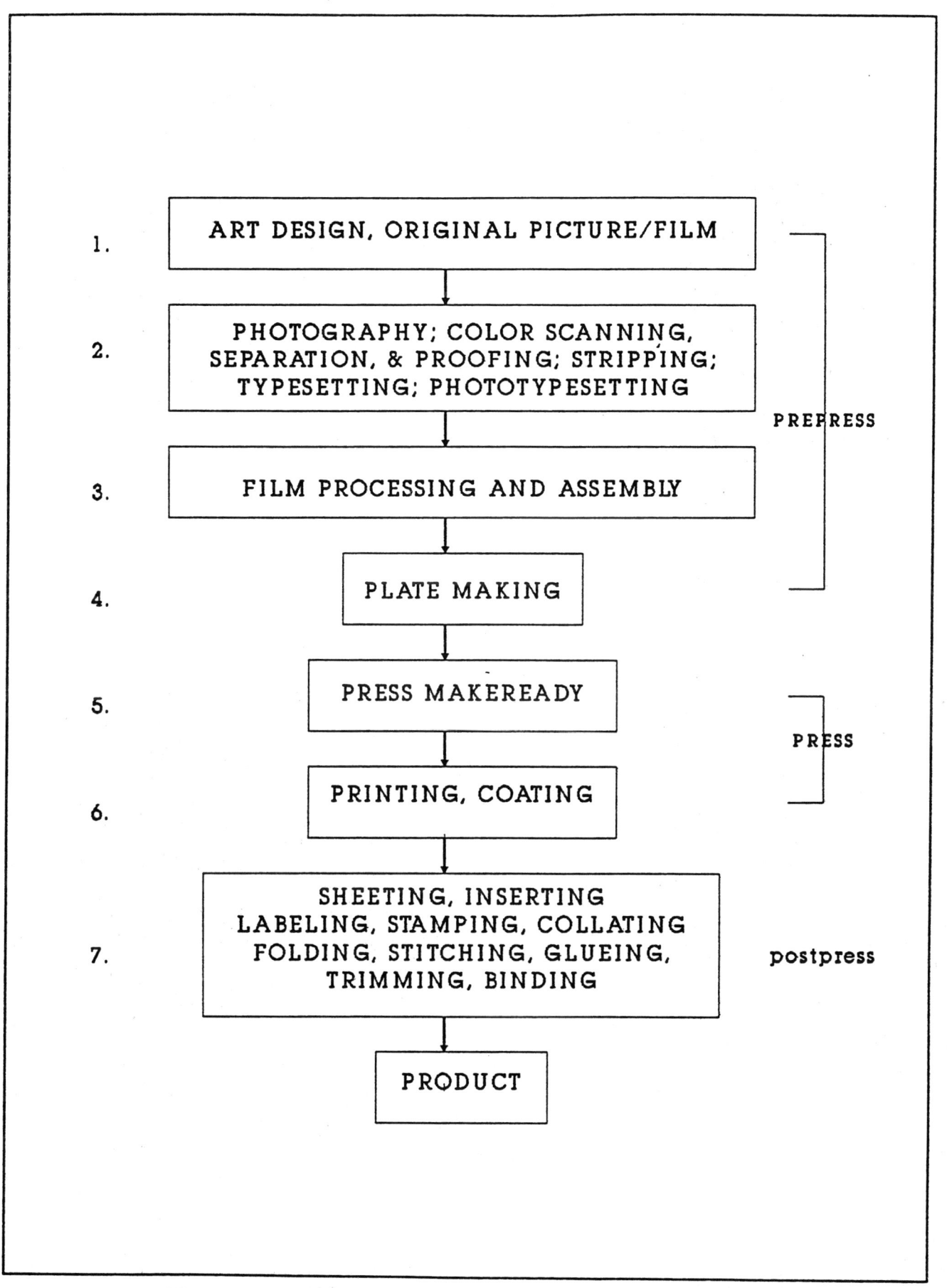

```
1.  ART DESIGN, ORIGINAL PICTURE/FILM
           │
2.  PHOTOGRAPHY; COLOR SCANNING,
    SEPARATION, & PROOFING; STRIPPING;
    TYPESETTING; PHOTOTYPESETTING
           │                                PREPRESS
3.  FILM PROCESSING AND ASSEMBLY
           │
4.  PLATE MAKING
           │
5.  PRESS MAKEREADY
           │                                PRESS
6.  PRINTING, COATING
           │
7.  SHEETING, INSERTING
    LABELING, STAMPING, COLLATING
    FOLDING, STITCHING, GLUEING,            postpress
    TRIMMING, BINDING
           │
    PRODUCT
```

Figure 3. Nonheatset Web Offset Operation Process Flow Diagram

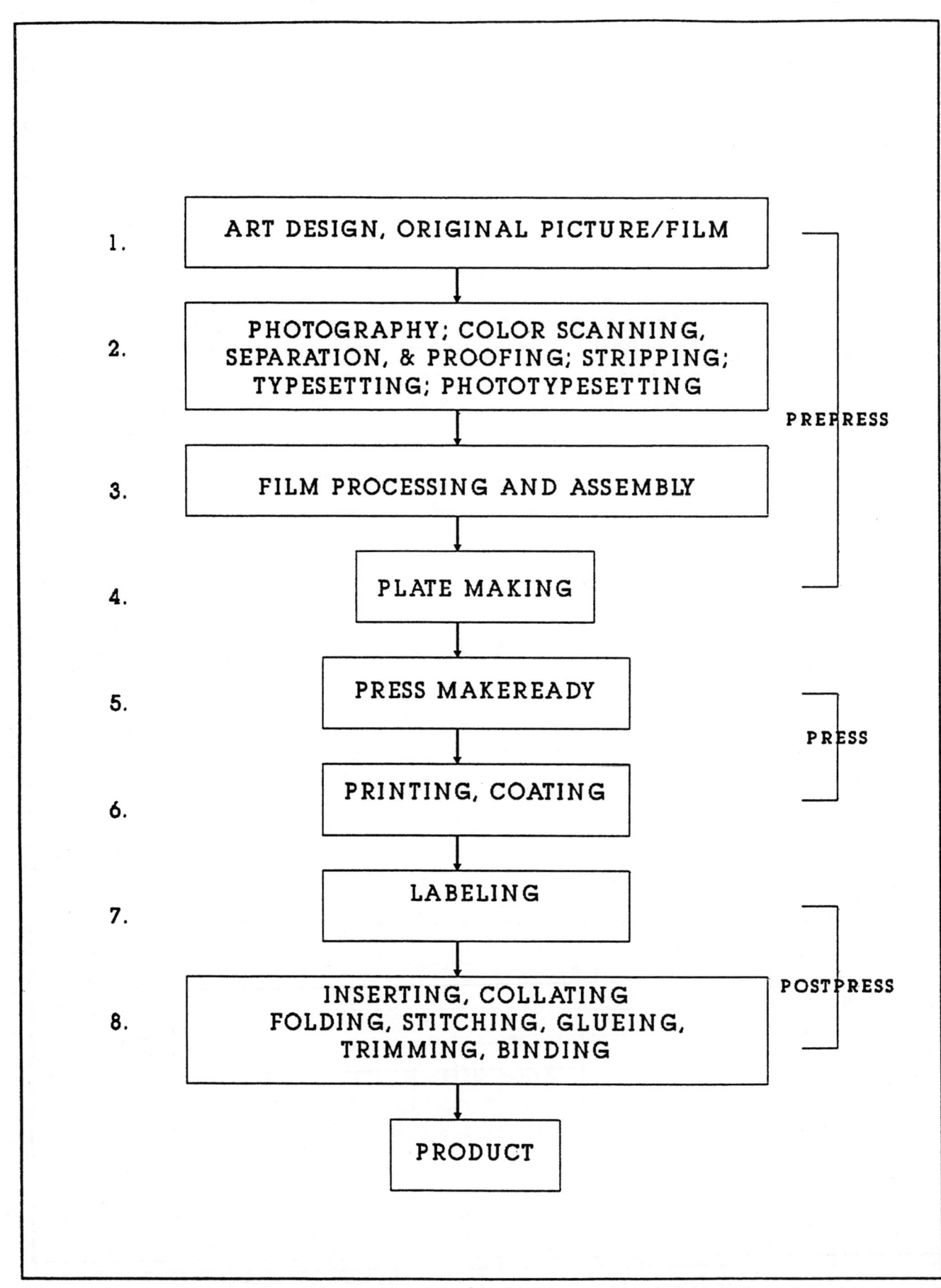

```
1.  ┌─────────────────────────────────────┐  ┐
    │   ART DESIGN, ORIGINAL PICTURE/FILM │  │
    └─────────────────────────────────────┘  │
                      ↓                       │
2.  ┌─────────────────────────────────────┐  │
    │   PHOTOGRAPHY; COLOR SCANNING,      │  │
    │   SEPARATION, & PROOFING; STRIPPING;│  │ PREPRESS
    │   TYPESETTING; PHOTOTYPESETTING     │  │
    └─────────────────────────────────────┘  │
                      ↓                       │
3.  ┌─────────────────────────────────────┐  │
    │     FILM PROCESSING AND ASSEMBLY    │  │
    └─────────────────────────────────────┘  │
                      ↓                       │
4.            ┌──────────────┐               │
              │ PLATE MAKING │               │
              └──────────────┘               ┘
                      ↓
5.          ┌──────────────────┐             ┐
            │  PRESS MAKEREADY │             │
            └──────────────────┘             │ PRESS
                      ↓                       │
6.          ┌──────────────────┐             │
            │ PRINTING, COATING│             │
            └──────────────────┘             ┘
                      ↓
7.            ┌──────────────┐               ┐
              │   LABELING   │               │
              └──────────────┘               │
                      ↓                       │ POSTPRESS
8.  ┌─────────────────────────────────────┐  │
    │    INSERTING, COLLATING             │  │
    │    FOLDING, STITCHING, GLUEING,     │  │
    │    TRIMMING, BINDING                │  │
    └─────────────────────────────────────┘  ┘
                      ↓
              ┌──────────────┐
              │   PRODUCT    │
              └──────────────┘
```

Figure 4. Flexography (Publication) Process Flow Diagram

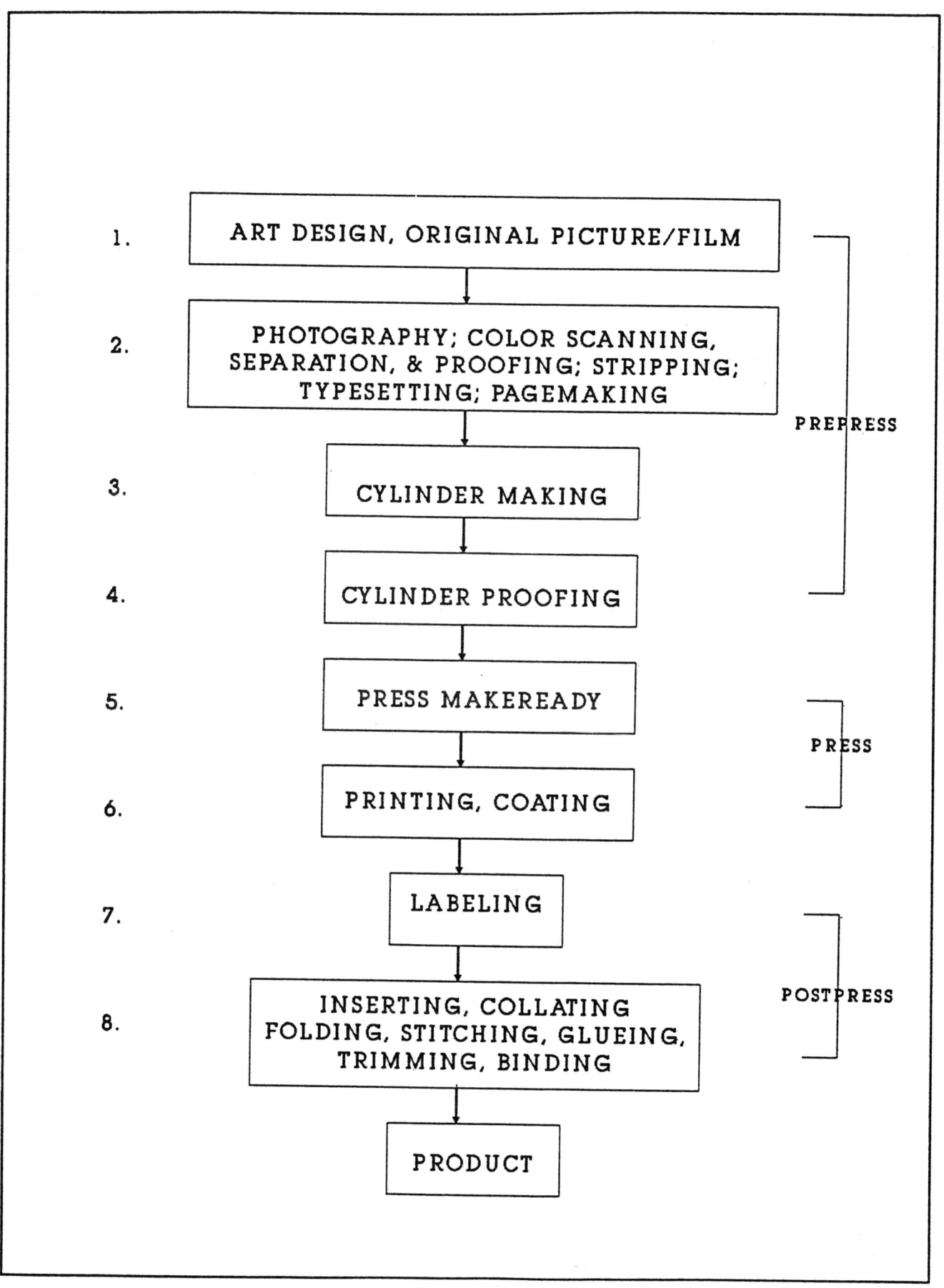

```
1.   ┌─────────────────────────────────────┐              ┐
     │  ART DESIGN, ORIGINAL PICTURE/FILM  │              │
     └─────────────────────────────────────┘              │
                        │                                 │
                        ▼                                 │
2.   ┌─────────────────────────────────────┐              │
     │     PHOTOGRAPHY; COLOR SCANNING,    │              │
     │   SEPARATION, & PROOFING; STRIPPING;│              │ PREPRESS
     │      TYPESETTING; PAGEMAKING        │              │
     └─────────────────────────────────────┘              │
                        │                                 │
                        ▼                                 │
3.         ┌─────────────────────┐                        │
           │   CYLINDER MAKING   │                        │
           └─────────────────────┘                        │
                        │                                 │
                        ▼                                 │
4.         ┌─────────────────────┐                        │
           │  CYLINDER PROOFING  │                        ┘
           └─────────────────────┘
                        │
                        ▼
5.         ┌─────────────────────┐                        ┐
           │   PRESS MAKEREADY   │                        │
           └─────────────────────┘                        │ PRESS
                        │                                 │
                        ▼                                 │
6.         ┌─────────────────────┐                        │
           │   PRINTING, COATING │                        ┘
           └─────────────────────┘
                        │
                        ▼
7.              ┌──────────────┐                          ┐
                │   LABELING   │                          │
                └──────────────┘                          │
                        │                                 │
                        ▼                                 │ POSTPRESS
8.   ┌─────────────────────────────────────┐              │
     │       INSERTING, COLLATING          │              │
     │  FOLDING, STITCHING, GLUEING,       │              │
     │         TRIMMING, BINDING           │              ┘
     └─────────────────────────────────────┘
                        │
                        ▼
                ┌──────────────┐
                │   PRODUCT    │
                └──────────────┘
```

Figure 5. Gravure (Publication) Printing Process Flow Diagram

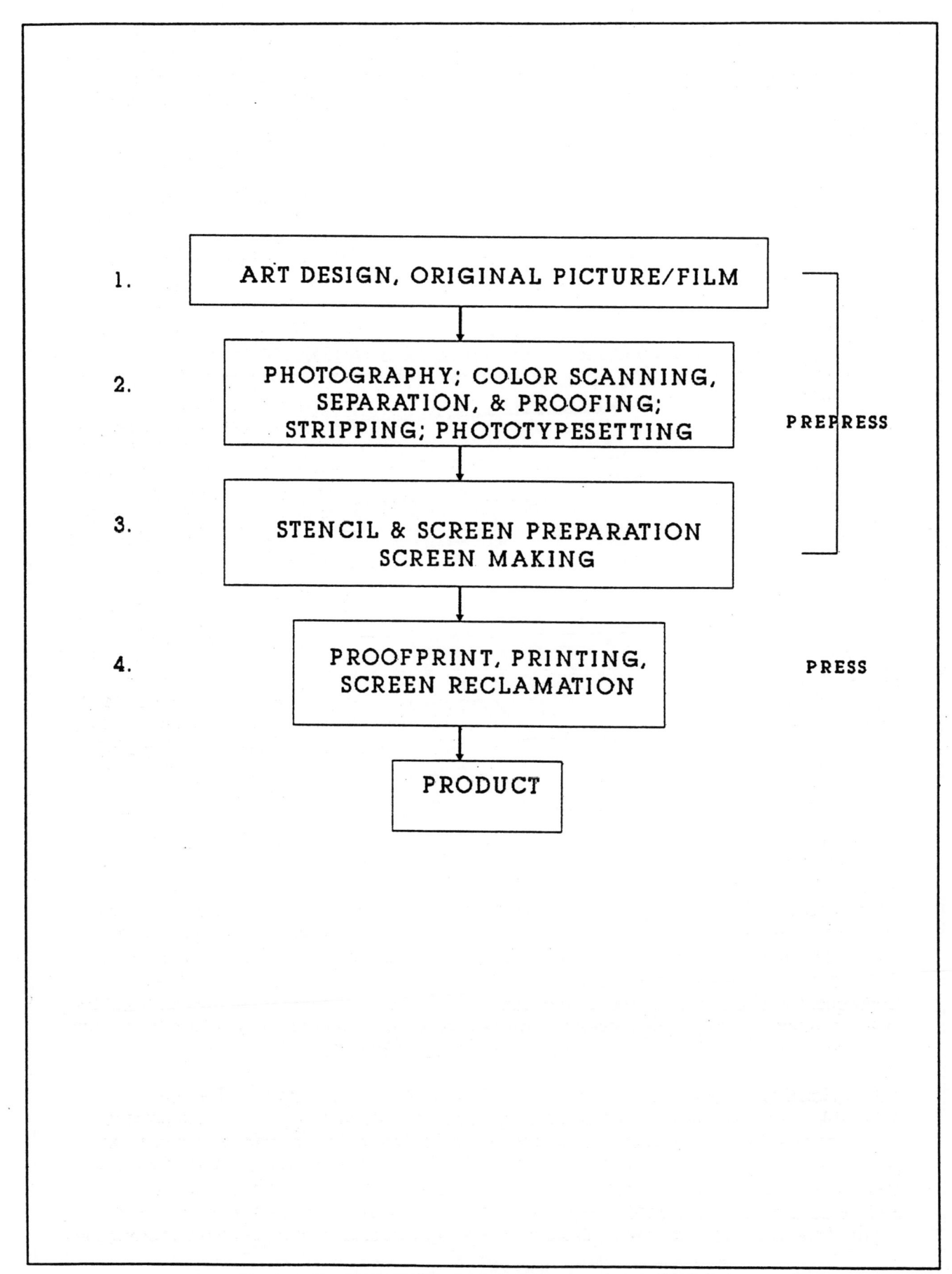

```
1.   ┌─────────────────────────────────────┐ ┐
     │  ART DESIGN, ORIGINAL PICTURE/FILM  │ │
     └─────────────────────────────────────┘ │
                        │                    │
2.   ┌─────────────────────────────────────┐ │
     │  PHOTOGRAPHY; COLOR SCANNING,       │ │
     │  SEPARATION, & PROOFING;            │ ├─ PREPRESS
     │  STRIPPING; PHOTOTYPESETTING        │ │
     └─────────────────────────────────────┘ │
                        │                    │
3.   ┌─────────────────────────────────────┐ │
     │  STENCIL & SCREEN PREPARATION       │ │
     │  SCREEN MAKING                      │ │
     └─────────────────────────────────────┘ ┘
                        │
4.   ┌─────────────────────────────────────┐
     │  PROOFPRINT, PRINTING,              │    PRESS
     │  SCREEN RECLAMATION                 │
     └─────────────────────────────────────┘
                        │
               ┌────────────────┐
               │    PRODUCT     │
               └────────────────┘
```

Figure 6. Screen Printing Operation Process Flow Diagram

flaw is made at the beginning of the process and not detected in time, it may carry through to the final product. The final product eventually will be rejected by the customer and a rerun is inevitable. This leads to the second prominent characteristic of printing the "chained and blown-up effect of waste generation" [3]. As shown in Fig. 7, the amount of waste generated in a typical printing process could pile up and expand like a balloon to an outrageous proportion if total quality control is not carefully instituted in the process. The wastes produced encompass not only streams generated during the operation but also everything that is used, produced, and involved in making the final product including the photoprocessing chemicals used; films and plates made; paper wasted; various inks applied; wash-up solvents used; and manpower, equipment time, and utilities utilized. In other words, much more waste and pollution can be generated in addition to those waste streams that can be physically evaluated. Some of these things cannot be physically analyzed (e.g., energy which is indirectly tied to air pollution) [2]. This probably is the single most influential factor that leads to a majority of the waste and pollution problems confronting most printing plants. Total quality management in a printing plant involves every process and employee. The reason is that each process is equally important. Every person involved takes mutual responsibility. Failure to recognize this unique situation in a printing plant impedes the reduction of waste and pollution to a reasonable level. Unfortunately, many U.S. printing plants do not exercise this very critical concept. Some printing plants generate paper waste at a rate as high as 50 percent [4]. If a unique total quality management concept could be developed for printers, a significant amount of waste including paper, ink, solvent, and various other pollutants could be reduced and, in cases, prevented. A quality makeready by well trained and educated pressmen is also essential to controlling waste generation and preventing pollution.

3.2. Makeready Streamlining. The makeready process precedes printing during color or job change. See Fig. 7, using lithographic printing as an example [3]. Makeready usually involves cleaning the entire roller trains, metering rollers, and blankets; changing and cleaning plates; changing inks and paper; replacing fountain solution; resetting print settings; and running the press. In the process, a large volume of inks, wash-up solvents, chemicals, and paper are used. The process is designed to bring the press up to maximum optimum printing conditions for production in which, hopefully, all flaws have been detected and eliminated. The quality control on makeready should not be based on the time required. Instead, print quality should dictate the duration of makeready. Certainly, fully understanding the quality process is essential to successfully controlling makeready.

A typical makeready can take 45 minutes to an hour on a typical multicolor commercial offset printing press. During makeready, everything is running including the press, ink, and paper. Yet, no product is made until the printed material is accepted by the customer or management as final product. At that time, the makeready is complete and the production process begins. In other words, during makeready, everything is wasted until production starts. The amount of waste paper generated during makeready varies from plant to plant. Some makeready processes run as high as 30 percent of the total paper purchased [4]. Consequently, if printers can reduce the amount of time each printing press uses on makeready or the number of makereadies performed on each press, they can definitely reduce the amount of waste generated and, at the same time, reduce associated pollution.

3.3. Printing Job Queue by Artificial Intelligence, An Expert System. The most noticeable makeready problem encountered by printers is scheduling jobs on each press to minimize numbers of makereadies necessary and associated time. In order to achieve that goal, printers must schedule printing jobs in queue on each press with some sort of ink color preference. As a result, pressmen need not to spend substantial time to ready the press for each printing job. Instead, they keep running the press without changing inks and perform any required and associated wash-ups, especially in plants that specialize in short-run printing jobs.

One of the easiest and quickest ways to accomplish this is to utilize artificial intelligence or an expert system to sequence the queuing of printing jobs on each press every day since most printers are well versed and equipped with personal computers [5]. An expert system can reduce the amount of time required to que jobs manually and also increase the accuracy of queuing. The main goal of such a system is to minimize makeready and, consequently, reduce the amount of wash solvent and ink used and the VOC emissions and hazardous wastes generated.

4. A Lubricating Oil Recovery System

Printers use large quantities of lubricating and hydraulic oils due to the various types of machinery (e.g., printing presses and bindery equipment) employed in their plants. Thus, significant quantities of waste oils are routinely generated. Those waste oils are normally disposed of through incineration either at a permitted facility or a cement kiln for energy recovery. Either way, the waste oils are not recovered for reuse. From an environmental point of view, the waste oils should be recovered for reuse since hydrocarbon oils are not renewable resources. A very simple device for separation and filtration should be able to segregate and clean the waste oil streams and allow reuse. This is a perfect way to minimize waste (i.e., minimizing oil wastes) and prevent pollution (i.e., preventing air pollution).

5. Lithographic Printing Ink Recycling and Reuse

Some lithographic printing inks (i.e., heatset and nonheatset web inks) recycling facilities have been established in the last few years. These facilities literally reclaim spent and used inks from printers, make black ink out of those waste inks, and, eventually, sell the black ink back to the printers at a much lower price. Only a minimum amount of reprocessing waste is generated. this type of reuse and recycling literally reduces the ultimate disposal of spent inks through incineration to nearly zero and, at the same time, reduces the amount of air pollution generated through incineration to almost zero [6]. Ink recycling is by far the most economic and environmentally sound approach to waste ink management.

6. Ongoing Pollution Prevention Initiatives in the Printing Industry

6.1. Spent Photoprocessing Chemical Recycling. Several large prepress chemical manufacturers have initiated recycling programs for spent photoprocessing chemicals, including those used in both film and plate making. Those that have done so aim to eliminate the need to discharge those spent processing chemicals into sewers and to recycle and reuse those chemicals [7].

6.2. Direct to Film and Direct to Plate Technologies. These technologies were originally developed to enhance the productivity and shorten the production time on prepress operation. These technologies have very distinctive benefits. The most prominent is that they eliminate the use of all photoprocessing chemicals in the prepress area. No pollution is actually generated when using those technologies during production [2].

TYPICAL LITHOGRAPHIC PRINTING OPERATION

CHAINED, IRREVERSIBLE PROCESSES

- ART DESIGN, ORIGINAL
- STRIPPING, TYPSETTING, PROOFING
- FILM-MAKING
- PLATE-MAKING
- MAKE-READY
- PRINTING, EMBOSSING, COATING, BRONZING, LAMINATING, STAMPING
- DIECUTTING, INSERTING, COLLATING, FOLDING, STITCHING, GLUEING, TRIMMING, BINDING
- PRODUCT

PREPRESS: Stripping/Typsetting/Proofing, Film-making, Plate-making
PRESS: Make-ready, Printing/Embossing/Coating/Bronzing/Laminating/Stamping
POSTPRESS: Diecutting/Inserting/Collating/Folding/Stitching/Glueing/Trimming/Binding

CHAINED BLOWN-UP EFFECT ON WASTE GENERATION

1. start
2. (circle)
3. (larger circle)
4. (largest circle)

○ → TOTAL AMOUNT OF WASTE GENERATED

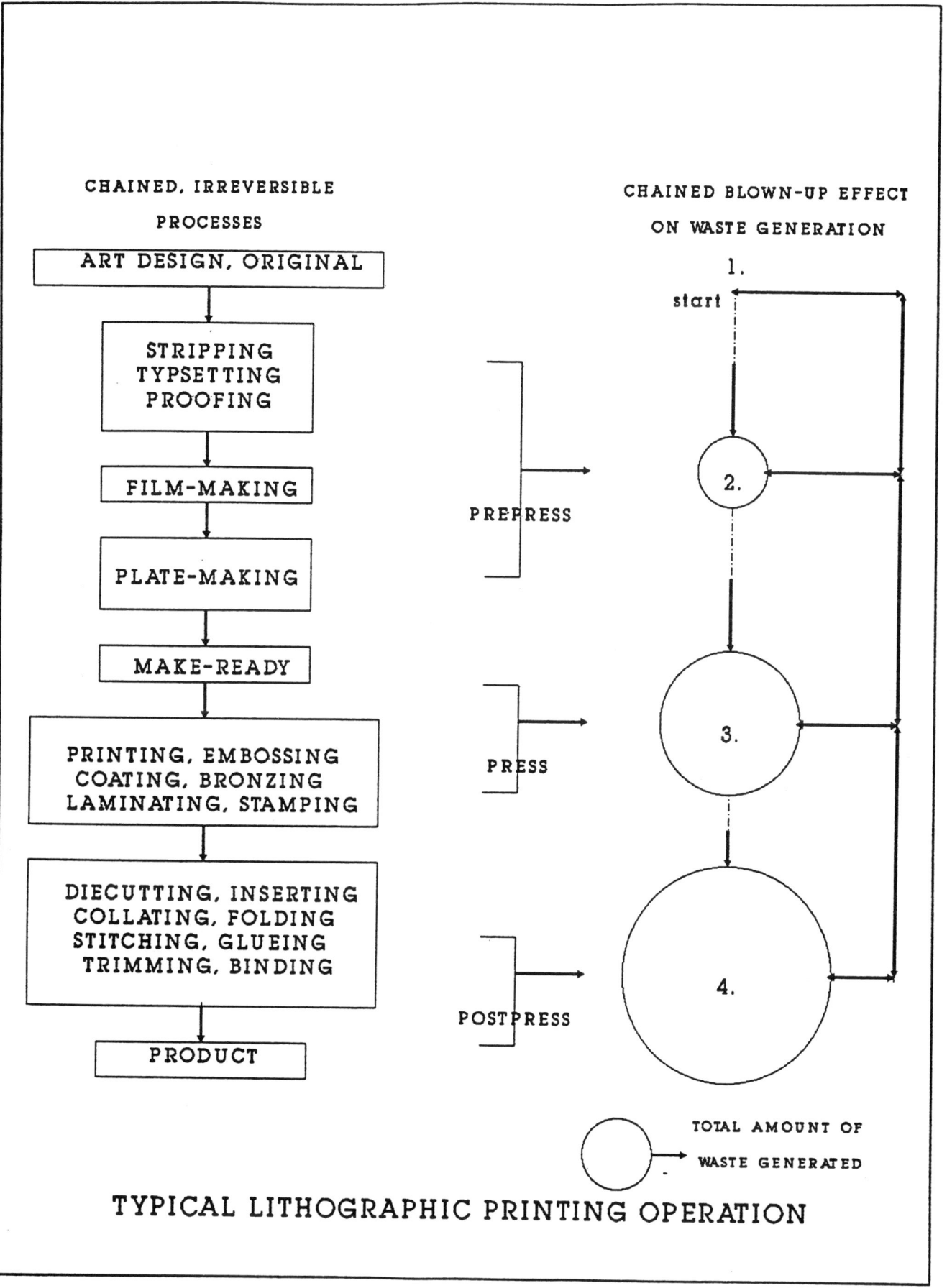

Figure 7. Chained, Blown-up Effect on Waste Generation

6.3. Water Washable Inks and Water Based Wash-up Solutions. This may sound impossible. However, one of the largest check printing companies in the U.S. recently announced that it had developed a new type of offset ink which can be washed and cleaned with a specially formulated water-based wash-up solution. If this technology could be confirmed, it would eliminate most VOC emissions from almost all existing printing plants by replacing all organic wash-up solvents with this solution [7].

6.4. Computerized Imaging and Printing Systems. Several new imaging and printing systems can bypass all traditional printing methodologies, namely film making, plate making, and printing [7]. These systems can take an image from a picture, manipulate it on the screen controlled by a computer, and have it printed on computerized printing equipment with specialized multi-color printing inks. These new systems literally eliminate all chemicals used in conventional printing and, thus, all associated pollution. Currently, those systems can only replace certain types of lithographic printing on a very small scale. Currently, however, these systems certainly have not reached the stage where they can replace all conventional printing [7].

7.   References

[1]   Michael Bruno, The Blue Book of the Printing Industry, 1993.

[2]   C. Nelson Ho, Pollution Prevention in The Printing Industry, Pittsburgh, PA: Graphic Arts Technical Foundation, July 1992.

[3]   C. Nelson Ho, Waste Minimization An Integrated Approach, 1992 TAGA Proceedings, p. 616, Rochester, NY: TAGA.

[4]   Robert Y. Chang and C. Nelson Ho, TQM as a Strategy for Waste Minimization, 1994 TAGA Conference, Baltimore, MD.

[5]   C. Nelson Ho, Pollution Prevention for Lithographic Printers through the Use of Artificial Intelligence An Expert System, 1991 Advances in Printing Science and Technology Proceedings, p. 28, London: Pentech Press.

[6]   Hope Gaines, The Three R's, The American Ink Maker, April 1994, pp. 24-30.

[7]   C. Nelson Ho, Personal information.

About the Author

C. Nelson Ho is the Manager of Corporate Environmental and Safety Services for the Graphics Arts Technical Foundation.

Pollution Prevention in the Textile Industry
Brent Smith
Professor of Textile Engineering, Chemistry, and Science
North Carolina State University
Raleigh, North Carolina 27695
(919) 515-6548

Pollution prevention for the textile industry is reviewed including present practices and future needs and opportunities. These are discussed generally as well as on a process-by-process basis.

Key Words

Dyeing; dyes; fabrics; fibers; finishing; textiles; pollution; pollution prevention; printing; source reduction; waste minimization

1. Introduction

1.1. Overview. The U.S. textile industry has a major impact not only on the nation's economy but also the economic and environmental quality of life in many communities. In several states, textiles is the leading provider of manufacturing jobs, with North Carolina as the nation's number one state in primary textile employment and production. There are about 1,000 textile manufacturing facilities in North Carolina, and about 6,000 nationwide. Apparel employment and production are led by New York, California, and North Carolina. The primary textile manufacturing industry provides nearly one fourth of total manufacturing employment in some states, and this number increases to well over one third if related industrial employment, (e.g., apparel manufacturing, fiber production, and textile machinery) is included. In 1992, the textile complex generated a gross national product of $72.5 billion, an amount that surpasses the automotive industry, primary metals, or petroleum refining [1].

Textile processing uses vast amounts of water, chemicals, and energy. Therefore, it strongly affects environmental quality in textile manufacturing regions [2,3]. In North Carolina alone, there are almost 200 NPDES-permitted textile wastewater point discharge sources, as well as hundreds of municipal systems in which textile operations are significant industrial users. A wastewater pollutant reduction of merely one part per million average across the textile industry in North Carolina alone is equivalent to roughly 150 tons per year of pollutant discharged to the environment. This indicates the immense potential of pollution source reduction.

Air quality issues are also important in textiles, with 132 facilities in North Carolina alone reporting air toxics emissions of over 3,000 tons in the aggregate [4]. In addition to air toxics concerns, most textile mills operate steam generation plants which produce boiler emissions. Although small in comparison to public utility electric power generating plants, textile sources are significant. In addition, the relationship of textiles to indoor air quality is an emerging issue now under intense scrutiny [5].

Solid waste is also produced in large quantities, and hazardous waste in small quantities in some operations. In many cases, destruction of these wastes is not feasible for economic, technical, or political reasons. For example, North Carolina ranks eleventh nationally in the production of hazardous waste, but has no permitted hazardous waste disposal facility within its borders.

All of the above illustrate the regional and national importance of pollution control in textile operations.

1.1.1. Textile Products, Markets, Applications. Textiles includes a wide variety of easily recognizable products, traditionally categorized into five groups: apparel, industrial, domestics, carpet/rug, and home furnishings. In addition, a broader view includes other important categories (e.g., nonwovens, films, fiber-reinforced composites, and nonfiberous polymers). The commercial importance of these nontraditional categories is increasing in products such as high-performance aircraft components, artificial arteries for surgical use, earthquake-proof bridge supports, and printed circuit boards. Of course, each end use product type requires different characteristics, which are generally provided by chemical and mechanical treatments.

All of the above textile products and millions more are made from a rather limited list of man-made and natural polymeric materials. Each type of fiber has different physical and chemical characteristics which make it amenable to certain end uses, and which also dictate different processing chemistry and machinery [7].

1.1.2. Textile Machinery and Processes. Textile processes can be broadly categorized as either batch or continuous. The type of processing and amounts of waste therefore depend not only on fiber content and intended end use of the textile product, but also on the processes and machinery selected. Process selection is in turn dictated not only by factors such as fiber and end use considerations as noted above, but also cost, production volume, physical form of substrate (e.g., fiber, yarn, knit, woven, nonwoven, garment), availability, scheduling, safety, and waste considerations. For example, cotton can be dyed with any one of seven dye classes: vat, sulfur, naphthol, pigment, direct, fiber reactive, or mordant. Then, within each class, hundreds of dyes are candidates. The selection of dyes, chemicals and machinery is critical in terms of the waste amounts and characteristics. A complete review of this subject is beyond the scope of this document, but further information is readily available [7].

1.2. State of the Art: Commercial Textile Practices. Textile waste amounts and characteristics are well documented as are state-of-the-art pollution prevention practices [2,3,8-14]. To summarize those documents, current commercial textile pollution prevention practices include material substitution, process modification, inventory control, better management techniques, recovery, and reuse. While several types of wastes have already been successfully targeted by the industry, four priority problem areas have been emphasized. These are: hard-to-treat wastes, discharged wastes, offensive or hazardous wastes, and large-volume wastes.

Hard-to-treat wastes, (i.e., wastes that are persistent or resist normal treatment) associated with textiles include color; metal; phosphate; phenol; and certain organic materials, especially surfactants, which resist biodegradation. Because of the extremely expensive and difficult procedures involved in removing these problem pollutants via wastewater treatment, source reduction is an economical and attractive alternative.

A major challenge of the textile industry is source reduction of waste which becomes widely dispersed when discharged. Wastes from textile processes often are reduced or captured for recycle/reuse by process modifications at the source because, once discharged, they tend to become widely dispersed and hard to treat. Machinery design, chemical substitution, procedure changes, or primary control measures can often accomplish better results and cost less than treatment. In addition, reclaimed waste in concentrated form (i.e., not dispersed) usually has its highest potential commercial reuse value.

Offensive or hazardous wastes, especially materials of high aquatic toxicity is a third problem. These wastes include metals, various types of organic solvents, and surfactants. In many instances, chemical substitutions can effectively reduce production of undesirable process by-products. Frequently, treatment of these hazardous or toxic process wastes leads to undesirable waste treatment solids (e.g., metal bearing sludges).

Large volume wastes are also a challenge. These wastes have successfully been reduced by process modification, chemical substitution, or on-site or off-site reuse. Each of the above types of waste may originate from a variety of textile operations.

The textile industry has an outstanding, documented record of pollution prevention activities [15]. Many case histories of reduction/conservation strategies, process modifications, chemical substitutions, and reclamation/reuse techniques which reduce these wastes in textile wet processing including preparation, dyeing, printing, finishing, and casual (miscellaneous) sources are available [15]. Case histories, in-plant techniques, and actual production experiences have generally resulted in economic gains for the processor as a separate benefit from the waste reduction, and have improved compliance with permits and/or pretreatment specifications.

The main purpose of this document is, however, to define specific and identifiable needs and opportunities for the textile industry to improve its already extensive efforts to prevent pollution.

2. Technical Needs and Opportunities for Pollution Prevention

Meeting several specific technical needs will increase the textile industry's ability to reduce pollution at its sources. Global needs, not assignable to any specific process, will be reviewed and opportunities on a process-by-process basis will be discussed. Needs of a nontechnical nature will be reviewed in Section 4.

2.1. General Needs. Overall technical needs and opportunities which conceptually should not be limited to any specific process include:

> Developing overall global views of pollution prevention
> Evaluating and applying known pollution reduction technologies
> Developing information systems to facilitate optimal decision making from design stage to consumer
> Improving process understanding

Unlike most industries, textiles is highly fragmented. Developing an overall global outlook is particularly difficult in textile manufacturing because each production unit's internal business is its primarily interest.

Global View. Most textile manufacturers have initiated pollution reduction on a process-by-process basis, but few have achieved the kind of global thinking which will maximize results. At each step, decisions are made which impact downstream processes beginning with product design, continuing through each process, and, ultimately, involving even the consumer. In many cases, processing assistants are added only to be removed later by energy and chemically intensive scouring procedures. Additives include lubricants, spin finishes, agricultural chemicals, size materials, knitting oils, winding lubricants, and tints. The challenge is to develop an overall global pollution prevention scheme which transcends individual operations. Such an outlook would consider the downstream impact of processing residues in terms of interferences and incompatibilities. This is discussed for specific processes in Sections 2.2.

To implement this global perspective requires better information exchange systems, which, in most cases, do not exist in the textile manufacturing complex. This information issue is discussed in more detail below. Another requirement is incentives, which can be difficult to establish in a fragmented manufacturing complex such as that used for textiles.

Global needs include not only the issues of design, additives' effects on downstream processing, and the need for later removal by scouring, but also developing an overall view of manufacturing in terms of containers, machine parts, print screens, drums, mix tanks, etc. to reduce solvent loss, drag out, and impurity build up. This affects not only process' pollution, but also product quality. Such thinking should extend even to consumer use of product in terms of aftermarket treatments, cleaning solvents, use conditions, installation, and maintenance while keeping in mind that the product itself will eventually become a waste.

Accurate Information. Application of known technologies based on documented studies often produces great benefits, however, accurate information about the pollution potential of various processes and products is needed to ensure optimal results. A surprising amount of confusing information is distributed, however, for two fundamental reasons. One is the lack of standardized environmental terminology and testing methods. The other is the propagation of obsolete data from old literature. The latter is a greater problem in textiles than other industries because, despite the fact that textiles is a mature industry, its chemistry has been developed relatively recently. The entire industry changed with the commercial development of synthetic fibers (e.g., polyester) in the 1950s and fiber reactive dyes for cotton in 1956. In addition, the entire chemistry of crosslinking resins has evolved since the 1930s, with modern finishes being developed almost entirely since 1970. Unfortunately, technical information developed prior to 1950 is often quoted in reviews in spite of excellent and easily accessible information sources [2,3]. Most of this obsolete information was developed during a time when commercial processing was not comparable to modern practices. A comprehensive and critically accurate literature review is needed for the textile mills. The opportunity is to accomplish great pollution reduction while at the same time increasing profits by using proven methods. The challenge is to critically review literature in terms of modern textile commercial practices as well as the best pollution prevention practices.

Information Distribution. Another major global need is to develop information distribution systems which will facilitate maximum pollution reduction practices. In essence, two types of information structures must be united. In textiles, there is a large gap between the management information systems and technical information systems. The challenge is to provide management, sales, and design personnel with information which enables proper decision making (e.g., product design, process selection, scheduling, and marketing). Although management and technical information must differ in format to accommodate differing backgrounds (i.e., less technical experience of management/sales/design vis a vis less business experience of chemists/engineers). It must have significant correlation and unification of technical and business concepts. Several attempts to develop such information systems with significant degrees of success.

Technical Understanding. At a fundamental level, even some of the most common and straightforward textile chemicals and processes are not well understood by experts, much less production supervisors, workers and managers. For example, conditions leading to unreacted monomer and catalyst in fiber, the process of bleaching, the role of spin finishes, surface phenomena (e.g., fouling), preparation processes, and dye aggregation are, in many cases, poorly understood. This is true for machines, fiber, chemical, and process issues. Other examples include the role of chlorinated solvents in cleaning; optimal methods for use of non-production chemical cleaners for dye machines, pad rolls, printing screens and rollers; the role of knitting oils and warp sizes; and how silicate stabilizes peroxide bleaching. The opportunity is to reduce pollution through better fundamental understanding of processes and systems.

Separation and Segregation. High-volume wastes have their maximum reuse potential prior to mixing with other wastes. Likewise, treatment of difficult wastes is often easier prior to mixing or dilution. Therefore, separation and segregation methods are broadly applicable for facilitating waste reuse on economic and technical grounds. The challenge is to study and optimize waste handling in terms of new facility design as well as retrofit applications.

2.1.1. Textile Wastes of Concern and Emerging Issues. As indicated in Section 1.2, the textile industry has developed the basis for a relatively comprehensive approach to pollution prevention by source reduction for several types of wastes. However, much remains to be done. Currently, the textile industry is being called upon to address even more difficult challenges in pollution control. Further efforts and resources will be required to solve the new important emerging environmental issues including:

- Indoor air quality
- Color residues in textile dyeing/printing wastewater
- Massive discharges of electrolytes
- Toxic air emissions
- Improving treatability of wastes
- Eliminating low levels of metals from wastewater
- Aquatic toxicity
- Continuing existing pollution reduction work

In addition to these new regulatory challenges, some very high volume waste streams deserve attention. These are salt, cutting room waste, knitting oils, and warp sizes.

One major technical challenge is to provide management, design, and planning departments with credible and convincing forecasts of future constraints and requirements based on technically accurate assessments of future regulatory issues such as those listed above. A critical study defining best pollution reduction management practices for specific future regulatory problems would provide a basis for developing better industrial pollution prevention programs.

It is not feasible to solve the above problems with treatment systems alone. In fact, improving waste treatment processes themselves often depends on producing more treatable, less dispersable, or less persistent wastes. Also, treatment is in many cases more efficient on concentrated waste. Therefore mixing together offensive or otherwise incompatible wastes is undesirable. Thus, future improvements in waste treatment are, in a sense, related to pollution prevention.

2.1.2. Reduction Strategies. Several pollution source reduction strategies are already known and widely used in textiles. These include common sense approaches such as:

>Design stage planning for processes
>Equipment maintenance and operations audit
>High extraction, low carryover process step separations
>Incoming raw material quality control
>Maintenance, cleaning, nonprocess chemical control
>Material utilization in cutting and sewing
>Optimized chemical handling practices
>Raw material prescreening prior to use
>Segregation, capture, recycle, reuse of wastes
>Training
>Developing a conservative worker attitude

Future improvements will use known but unused technologies (e.g., technology transfer from other industries) new technologies based on known science, and perhaps even the development of new science. Examples include:

>Better risk assessment methods, data, and procedures
>Better informed consumers, designers, management, suppliers
>Disposal facilities for captured waste
>Even better optimized chemical handling practices
>Higher purity raw materials
>Improved waste audit procedures
>Improved standard test methods and definitions
>Less disinformation and politics
>More global and integrated view of manufacturing
>More technology transfer
>More recycle opportunities
>More markets for wastes
>More chemical expertise and general industry competence

In Section 2.2 as well as Section 3, the above will be discussed on a process-by-process basis to define specific strategic needs, to describe opportunities for innovation and improvement, and to identify challenges and barriers to the fulfillment of these needs.

2.1.3. Engineering Practices. There are opportunities to apply certain engineering practices more broadly in textile manufacturing. These include the using engineering-based product design, developing emissions factors, and establishing standardized tests and nomenclature.

Engineering-Based Product Design. One particularly important need is to introduce engineering considerations into fabric design and production scheduling. Presently, textile fabric design is primarily coloristic, artistic, and aesthetic. Production scheduling is usually market driven with secondary consideration of problems created, for example, by excessive color changes and associated machine cleaning. Better design systems based on engineering and chemical principles will improve environmental and other process considerations. Designers and scheduling departments are rarely aware of product attributes (e.g., colors) which produce high pollution loads, cause scheduling difficulties, or require excessive machine cleaning. The opportunity is to provide artificial intelligence or expert systems which will assist designers and process schedulers. Examples include not only color selection, but also fiber blend selection, knit or woven constructions, etc. The same reasoning applies to sales/customer relationships which should, but often do not, consider the environmental impact of product specifications.

Emissions Factors. Advancements in engineering methodology are also needed in risk assessment and waste audits. Most mills use either direct measurement of waste for existing processes or mass balance estimates at the design stage to determine potential wastes from processes. Direct measurement of waste is often difficult, and, of course, cannot be done at the design stage before process start up. Mass balance estimates are severely limited in that the difference between large volume of raw materials and large volume of product leaves a small difference for the waste number, which (being the difference of two large numbers) is highly uncertain. Thus, there is a need for standard emissions factors for textile operations.

Using emission factors to predict the ultimate fate of pollutants can greatly improve waste audit accuracy. Also, the simplicity of using standard emissions factors provides an opportunity to examine processes more efficiently, and thereby reduce more pollution with a given amount of resources (e.g., time, personnel). This also offers an opportunity to better evaluate processes in terms of problem wastes and to target high-volume wastes from inefficient processes (e.g., garment dyeing) as well as difficult waste from efficient processes (e.g., flame retardant backcoating). Finally, emissions factors are one of the few accurate ways to predict trace and fugitive emissions such as hydrocarbons and metals.

Textiles, unlike most industries, has no standard emissions factors for many specific raw materials and processes. Two notable studies estimate amounts of certain pollutants based on production volume as shown in Table 1, but the pollutants included in that data are very limited [2,17]. Since these studies were completed, priority pollutants have been further regulated as have locally regulated organic compounds and toxic air pollutants. The need is to understand the environmental fate of all process chemicals, to identify the precursors of chemical wastes, and to develop emissions factors for various production situations. Then, process engineers and schedulers will have the opportunity to control these pollutants at the planning stage. Emissions factors are key to planning because many of the most offensive toxic air and water pollutants from textiles either are fugitive emissions or result from trace impurities in high-volume raw materials. In this case, mass balance is essentially useless in any practical sense. important that these studies include maintenance and machine cleaning chemicals as well as process chemicals.

Standard Tests and Nomenclature. Another need is for engineering definitions and standard test protocol of waste and environmental parameters. Currently, even the term "biodegradable" is not defined. Well-defined, quantitative understandable ratings for environmental and waste attributes (e.g., aquatic toxicity, treatability) will enable engineers to better assess risk, process trade offs, and the like.

Table 1. Water Pollutants Based on Production Volumes [2]

Subcategory	Water Usage, ℓ/kg Min.		(gal./lb.) of production Med.		Max.		Discharge cu m/day (MGD) Median Mill		No. of Mills
1. Wool souring	4.2	(0.5)	11.7	(1.4)	77.6	(9.3)	103	(0.051)	12
2. Wool finishing	110.9	(13.3)	283.6	(34.1)	657.2	(78.9)	1892	(0.500)	15
3. Low water use processing	0.8	(0.1)	9.2	(1.1)	140.1	(16.8)	231	(0.061)	13
4. Woven fabric finishing									
a. Simple processing	12.5	(1.5)	78.4	(9.4)	275.2	(33.1)	636	(0.168)	48
b. Complex processing	10.8	(1.3)	86.7	(10.4)	276.9	(33.2)	1533	(0.405)	39
c. Complex processing plus desizing	5.0	(0.6)	113.4	(13.6)	507.9	(60.9)	636	(0.168)	50
5. Knit fabric finishing									
a. Simple processing	8.3	(0.9)	135.9	(16.3)	392.8	(47.2)	1514	(0.400)	71
b. Complex processing	20.0	(2.4)	83.4	(10.0)	377.8	(45.2)	1998	(0.528)	35
c. Hosiery products	5.8	(0.7)	69.2	(8.3)	289.4	(34.8)	178	(0.047)	57
6. Carpet finishing	8.3	(1.0)	46.7	(5.6)	162.6	(19.5)	1590	(0.420)	37
7. Stock and yarn finishing	3.3	(0.4)	100.1	(12.0)	557.1	(66.9)	961	(0.254)	116
8. Nonwoven manufacturing	2.5	(0.3)	40.0	(4.8)	82.6	(9.9)	389	(0.100)	11
9. Felted fabric processing	33.4	(4.0)	212.7	(25.5)	930.7	(111.8)	564	(0.149)	11

Table 1 (Continued)

Subcategory	BOD	COD (kg/kkg)	TSS	O&G	Phenol	Chromium (g/kkg)	Sulfide
1. Wool scouring	41.8	128.9	43.1	10.3			
2. Wool finishing	59.8	204.8	17.2				
3. Low water use processing	2.3	14.5	1.6				
4. Woven fabric finishing							
a. Simple processing	22.6	92.4	8.0	9.1	8.2	4.3	7.6
b. Complex processing	32.7	110.6	9.6	3.8	7.7	2.6	12.5
c. Complex processing plus desizing	45.1	122.6	14.8	4.1	13.1	20.9	
5. Knit fabric finishing							
a. Simple processing	27.7	81.1	6.3	4.0	8.7	7.8	13.0
b. Complex processing	22.1	115.4	6.9	3.5	12.0	4.7	14.0
c. Hosiery products	26.4	89.4	6.7	6.6	4.2	6.4	23.8
6. Carpet finishing	25.6	82.3	4.7	1.1	11.3	3.4	9.4
7. Stock and yarn finishing	20.7	62.7	4.6	1.6	15.0	12.0	27.8
8. Nonwoven manufacturing	6.7	38.4	2.2			0.5	
9. Felted fabric processing	70.2	186.0	64.1	11.2	247.4		

Insufficient data to report value.

Specific Opportunities for Unit Processes. The previous information defines global needs and opportunities which can be applied to specific processing situations as those indicated in the following sections.

2.2.1. Fibers: Synthetic and Natural. There are several needs and opportunities from the fiber perspective. For synthetics, these include the development of improved spin finishes and fibers with lower amounts of residual monomer and catalyst. For natural fibers needs and opportunities include elimination of minerals, metals, and agricultural residues including biocides. Also, better fiber selection could afford the opportunity to reduce the amount of chemical finishing required.

Spin Finishes. Proprietary spin finishes are added to synthetic fibers to provide fiber lubrication and other desirable properties, such as static electricity control. The chemical composition of spin finishes is a closely guarded trade secret but, in almost all cases, they must be removed prior to dyeing and finishing to ensure uniform penetration of fabric by dyes and finishes, and to avoid reaction or precipitation with incompatible downstream process chemicals. Also, volatile components of these spin finishes produce air pollution when vaporized by high-temperature processes such as heatsetting, dye thermofixation, drying and curing of finishes. To prevent these problems, spin finishes must be scoured from the goods prior to dyeing and finishing, thereby producing polluted wastewater. Thus, a need exists for better ways to control the surface and electrical properties of fibers without the use of additives which interfere with downstream processing, quality, and associated pollution potential. The challenge is first to better understand the role of fiber lubricity in textile processes and textile structures, and then to use that knowledge to develop better spin finishes, which do not interfere or pollute. Desirable lubricity and electrical properties could potentially be provided by polymer surface modifications, such as introducing permanent additives, or by physico-mechanical treatment (e.g., plasma).

High Purity Fibers. Another challenge is to produce synthetic fibers with minimum amounts of unreacted monomer, and to use less harmful catalysts and additives (e.g., delusterants) in polymerization reactions. Many toxic pollutants from textile wet processing operations have been identified [8]. They are, in many cases, the same materials which can be extracted from raw fibers [18]. Making polymerization reactions more efficient and more robust to process variations potentially would reduce or eliminate these undesirable pollutants from wastewater streams of wet processors. Table 2 shows several examples of such pollutants. This same idea is discussed further in Section 2.2.4, "Chemical Commodities."

Natural fibers present a similar challenge, but involve different pollution issues. Agricultural residues such as pesticides, herbicides, and defoliants can lead to aquatic toxicity and other problems from preparation wastewater. Data of Table 3 show that processing solutions tend to leach metals out of textile substrates. The challenge is to develop cleaner agricultural production practices, to inherently increase insect and disease resistance, or to require less chemical additives, or perhaps safer or less offensive additives.

One trend of note with regard to impurities in fibers is the present fad of naturally colored cottons. These are grown with inherent colors, resulting presumably from the sorption of minerals from the soil during the growth process, or development of color from some biochemical process in the plant itself. To date, no data have been published showing the pollution potential of these fibers during scouring operations, nor the effects of these colorants

Table 2. Toxic Pollutants Found in Textile Wastewater [8]

The following organic chemical substances were identified in textile wet processing wastewater. Abbreviations as follows are used:

Exposure routes:

ihl	inhalation
orl	ingestion
unk	unknown exposure
eye	in the eye
skn	on the skin
ipr	in the peritoneal cavity
inv	in the vein - hypodermic

Exposure times:

M	minutes
H	hours
D	days
W	weeks
Y	years

Species:

hmn	human
man	man
inf	human infant
wmn	woman
rat	rat
cat	cat
mus	mouse
rbt	rabbit

Effects:

TDLo	Any nonlethal effect, all exposures but ihl
TCLo	Any nonlethal effect, exposure by ihl only
LDLo	Death, all exposures by ihl
LD50	Death of 50% of subjects, all exposures but ihl
LCLo	Death, exposure by ihl
LC50	Death of 50% of subjects, exposure by ihl

Toxicity data are reported for each chemical, if found. The exposure route, species, and effect are reported. If no effect is reported, then the test showed unspecified toxic effects. The last entry on the line (T...) is the reference number. All references are listed at the end of this section.

The carcinogenic determination for each, of available, is listed. The aquatic toxicity for each chemical is listed as "TLm96" which is the concentration of chemical which will kill 50% of the exposed organisms within 96 hours. Because of the lack of standard flow conditions (static, flow through) and variation of species used in the tests, the toxic concentrations are reported as ranges.

Table 2. Continued

Toxicity Data for Compounds

1,2 Benzenecarboxylic acid, butyl phenylmethyl ester
(Butyl phenylmethyl phthalate)

    no toxicity data were located

1,2 Benzenecarboxylic acid, dibutyl ester
(Dibutyl phthalate)

orl	rat	LD50 12000 mg/kg
orl	rat	TDLo 8400 µg/kg
ihl	rat	LC50 7900 µg/m3
orl	hum	TDLo 140 mg/kg Affects the eyes

    Aquatic Toxicity TLm96 1000-10 ppm

1,2 Benzenecarboxylic acid, decyl hexyl ester
(Decyl hexyl phthalate)

orl	rat	LD50 49 gm/kg

1,2 Benzenecarboxylic acid, diiosdecyl ester
(Diiosdecyl phthalate)

skn	rbt	10 mg (20H) Mild reaction
skn	rbt	LD50 17 g/kg

1,2 Benzenecarboxylic acid, ethyl hexyl ester
(Ethyl hexyl phthalate)

orl	man	TDLo 143 mg/kg
orl	rat	LD50 31 gm/kg
orl	rat	TDLo 35 mg/kg

Benzene, 1 methyl 4-(1-methylethyl)
(p-Cymene)

orl	rat	LD50 4750 mg/kg
ihl	rat	LCLo 5000 ppm (45M)

Benzene, trichloro

skn	rbt	1950 mg (13W) Moderate effects
orl	rat	LD50 756 mg/kg
orl	mus	LD 756 mg/kg

    Aquatic Toxicity 10-1 ppm

Table 2. Continued

Butyl cyclohexane

    no toxicity data were located, however, based on analogous compounds, toxicity is very low (skn TDLo > 25 g/kg)

Butyl ethyl cyclopentane

    no toxicity data were located, however based on analogous compounds, toxicity is very low (skn TDLo > 25 g/kg)

Camphene

    toxicity not reported

Camphor

orl	inf	LDLo 70 mg/kg
unk	man	LDLo 29 mg/kg
orl	rbt	LDLo 2000 mg/kg

Chlorocyclohexane

| ipr | mus | LD50 830 mg/kg |

Chloroform

orl	rat	LD50 800 mg/kg
inh	rat	TCLo 1000 ppm (7H)
orl	rat	TDLo 70 g/kg
orl	hum	LDLo 140 mg/kg
ihl	hum	TCLo 5000 mg/m3

Aquatic Toxicity TLm96 100-10 ppm
Animal positive carcinogen

Cyclohexane

orl	rat	LD50 29820 mg/kg
eye	hum	5 ppm
orl	rbt	LDLo 5500 mg/kg

Aquatic Toxicity TLm96 100-10 ppm

Cyclohexanol

    no numerical toxicity data were located
    "Moderately toxic via oral and inhalation routes"

Cyclohexanol, 2-chloro
ipr   mus   LD50 830 mg/kg

Table 2. Continued

Cycloheaxnone

    inh   hum   LCLo 75 ppm Irritant effects
    orl   rat   LD50 1620 mg.kg
    ihl   rat   LCLo 2000 (4H)

Aquatic Toxicity TLm96 100-10 ppm

Cyclopentane

    ihl   mus   LCLo 110000 mg/m3

Aquatic Toxicity TLm96 > 1000 ppm

p-Cymene

    orl   rat   LD50 4750 mg/kg
    ihl   rat   LCLo 5000 ppm (45M)

d-Limonene

    ipr   mus   TDLo 4800 mg/kg
    orl   mus   TDLo 67 g/kg

Decahydronaphthaline

    orl   rat   LD50 4170 mg/kg
    ihl   rat   LC50 710 ppm (4H)
    ihl   hum   TCLo 100 ppm (Irritant effects)

Aquatic Toxicity TLm96 1000-100 ppm

Decane

    skn   mus   25g/kg (1Y)

Dichloromethane
(Methylene chloride)

    ihl   rat   TCLo 4500 ppm (24H)
    ihl   hum   TCLo 500 ppm (1Y)
    orl   rat   LD50 167 mg/kg
    ihl   rat   LC50 88000 mg/m3 (30M)

Aquatic Toxicity TLm96 1000-100 ppm
Carcinogenic determination is suspect (indef)

Decosane no toxicity data were located, however, based on analogous compounds, toxicity is very low (skn TDLo > 25 g/kg)

Table 2. Continued

Docosane, 11 butyl

    no toxicity data were located, however, based on analogous compounds, toxicity is very low (skn TDLo > 25 g/kg

Dodecanamide, N,N, bis (2-hydroxyethyl)

    no toxicity data were located

Dodecan

    skn   mus   TDLo 11g/kg (22W)

Ethane, thiobis

    no toxicity data were located

Ethyl benzene

    orl   rat   LD50 3500 mg/kg
    ihl   rat   LDLo ppm 4000 (4H)
    ihl   rat   TCLo 97 ppm (7H)
    ihl   hmn   TCLo 100 ppm (8H)

    Aquatic Toxicity TLm96 100-10 ppm

Ethyl cyclohexane

    no toxicity data were located, however, based on analogous compounds, toxicity is very low (skn TDLo > 25 g/kg)

2-Ethyl Hexanol

    orl   rat   LD50 3200 mg/kg
    orl   msu   LDLo 3200 mg/kg

2-Ethyl hexyl adipate

    orl   rat   LD50 9110 mg/kg
    ivn   rat   LD50 900 mg/kg
    skn   rbt   16/kg

Heptacosane

    no toxicity data were located however, based on analogous compounds, toxicity is very low (skn TDLo > 25 g/kg)

Hexadecane

    no toxicity data were located however, based on analogous compounds, toxicity is very low (skn TDLo > 25 g/kg)

Hexanedioic acid

    skn   hum   75 mg mild reaction reported

Hexanoic acid, dioctyl ester

    no toxicity data located

Hexanoic acid, mono (2-ethyl hexyl) ester
(Ethyl hexyl adipate)

    orl   rat   LD50 9110 mg/kg
    ivn   rat   LD50 900 mg/kg
    skn   rbt   16/kg

Hexyl cyclohexane

    no toxicity data were located, however, based on analogous compounds, toxicity is very low (skn TDLo > 25 g/kg)

Isocineole

    no toxicity data were located

Lucenin 2

    no toxicity data were located

m-xylene

    see xylene

p-mentha-1,4-diene

    skn   rbt   500 mg Moderate effects
    orl   rat   LD50 3650 mg/kg

Methylene chloride
(Methane, dichloro)

    ihl   rat   TCLo 4500 ppm (24H)
    ihl   hum   TCLo 500 ppm (1Y)
    orl   rat   LD50 167 mg/kg
    ihl   rat   LC50 88000 mg/m3 (30M)

Table 2. Continued

Aquatic Toxicity TLm96 1000-100 ppm
Carcinogenic determination is suspect (indef)

Methylethyl cyclohexane

no toxicity data were located, however, based on analogous compounds, toxicity is very low (skn TDLo > 25 g/kg)

Methyldecane

no toxicity data were located, however, based on analogous compounds, toxicity is very low (skn TDLo > 25 g/kg)

Methyl ethyl benzene

orl   rat   LD50 5000 mg/kg

1-Methyl-4-isopropyl cyclohexadiene-1,3

orl   rat   LD50 1680 mg/kg

Methyl n-propyl cyclohexane

no toxicity data were located, however, based on analogous compounds, toxicity is very low (skn TDLo > 25 g/kg)

Myrcene (beta)

no toxicity data were located

Nonane

ivn   mus   LD50 218 mg/kg

o-xylene

see xylene

p-xylene

see xylene

Palmitic acid
(Hexanedioic acid)

skn   hum   75 mg Mild reaction

Pentacosane no toxicity data were located, however, based on analogous compounds, toxicity is very low (skn TDLo > 25 g/kg)

157

Table 2. Continued

Perchloroethylene
(1,1,2,2 Tetrachloroethylene)

orl	rat	LD50 8850 mg/kg
ihl	rat	LCLo 4000 ppm (4H)
ihl	rat	TCLo 1000 ppm (24H)
ihl	hum	TCLo 96 ppm (7H) Systemic effects

Aquatic Toxicity TLm96 100-10 ppm
Carcinogenic Determination - Animal Positive

Phthalate esters ... (all)

see 1,2 Benzenecarboxylic acid ...

Pinene (alpha)

skn	man	500 mg Severe effects
orl	rat	LD50 3700 mg/kg
ihl	rat	LCLo 625 μg/m3

Propanone (2)
(Acetone)

ihl	man	TDLo 440 μg/m3 (6M)
ihl	man	TDLo 10 mg/m3 (6H)
ihl	man	TCLo 12000 ppm Central nervous eff
orl	rat	LD50 9750 mg/kg
ihl	rat	LCLo 64000 ppm (4H)

Aquatic Toxicity > 1000 ppm

Propyl cyclohexane

no toxicity data were located

Stoddard Solvent

eye	hum	470 ppm (15M)
ihl	cat	LCLo 10 gm/m3 (2.5H)

Toluene

orl	rat	LD50 5000 mg/kg
ihl	rat	TDLo 1000 mg/m3
ihl	hum	TCLo 200 ppm

Aquatic Toxicity TLm96 100-10 ppm

Table 2. Continued

Trichlorobenzene

    skn   rbt   1950 mg (13W) Moderate effects
    orl   rat   LD50 756 mg/kg
    orl   mus   LD50 766 mg/kg

Aquatic Toxicity 10-1 ppm

Tricosane

no toxicity data were located, however, based on analogous compounds, toxicity is very low (skn TDLo > 25 g/kg)

Trimethyl benzene

    ihl   rat   LC50 18 mg/m3 (4H)
    orl   rat   LDLo 5000 mg/kg

Trimethyl cyclohexane

no toxicity data were located, however, based on analogous compounds, toxicity is very low (skn TDLo > 25 g/kg)

Terpinene (gamma)

no toxicity data were located

Undecane

    ivn   mus   LD50 517 mg/kg

Xylene

    ihl   rat   LC50 5000 ppm (4H)
    ihl   rat   TCLo 1000 mg/m3 (24H)
    ihl   hum   TCLo 200 ppm (Irritant effects)
    ihl   hum   LCLo 10000 ppm (6H)

Aquatic Toxicity TLm96 100-10 ppm

Table 3. Metals in Raw Water and Processing Solutions [30]

Raw Water Quality in Textile Mills in Southeastern USA

Constituent	Equivalency	Average of 10	Range lowest	highest
Calcium	CaCo$_3$	12.9	1.0	46.5
Magnesium	CaCo$_3$	3.8	1.5	7.8
Sodium	CaCo$_3$	36.0	5.7	76.1
Alkalinity				
Bicarbonate	CaCo$_3$	27.7	10.0	110.0
Carbonate	CaCo$_3$	1.4	0.0	10.0
pH	—	7.2	5.7	7.8
Iron	Fe++	0.1	0.01	0.31
Copper	Cu++	0.02	0.01	0.10
Manganese	Mn++	0.01	0.0	0.05
Zinc	ZN++	0.11	0.0	0.24

Contaminants in Processing Solutions in Textile Mills
Peroxide saturator solutions from three mills
bleaching cotton fabric in J boxes

Metal	Average of 14[3]	Dissolved metal in solution (ppm) Range lowest	highest
Ca	68.0	28.0	130.0
Mg	24.0	7.7	49.0
Fe	1.5	0.5	3.0
Cu	0.25	0.065	0.68
Mn	0.03	0.01	0.06
Zn	0.49	0.14	0.80

on the performance of products with regard to important safety issues (e.g., flammability). Further study is needed.

Better Fiber Performance. Another less well defined challenge is to utilize fibers which inherently require less chemical processing to achieve desired end use performance. In some cases, this is a design issue. In other cases, it is a fiber issue. This goes hand in hand with the concepts of improved product design practices, based on better fiber information for fabric designers. In view of the fact that most fabric designers are artistic rather than material science and engineering oriented, there is a need to better utilize fiber data (i.e., strength, surface characteristics, dyeability, reactivity, etc.) to optimize product design. Methods, perhaps artificial intelligence systems, to facilitate the design of products with better end use performance but requiring less chemical and other processing and aftercare requirements must be developed. This is extremely important in terms of indoor air quality issues, and is currently the subject of major research efforts.

2.2.2. Dyes. The first synthetic textile colorant was produced in the 1860s when Perkin oxidized aniline to produce Mauvine. By the early 1880's, diazotization was a known reaction, and chemists like Greiss, Walter, and Boettiger attempted to synthesize commercially useful dyes with that method. During the ensuing century of dye development, thousands of synthetic colorants have been produced and used commercially. Traditionally, dye manufacturers' goals have been to produce low-cost dyes with high tinctorial value, brilliance, and good application and fastness properties; in particular, high resistance to washing, crocking, light, oxidation (ozone), reduction (gas fading), chlorine attack, acids and alkalis, etc. Although safety has, over time, become a consideration in the synthesis of dye from intermediates, treatability has not been a significant consideration in dye design. Dye research, until now, has focused on dyes with improved stability and, thus, more resistance to treatment. For example, in the 1880s, dyes would fade in 5 standard fading units (SFU) of light exposure. By 1980, 50 to 100 SFU light was the norm. The next generation of dyes under current development for automotive uses will withstand over 1000 SFU. Chemists, by their success at developing highly stable dyes, have produced color wastes from dyeing and printing operations that are very difficult to treat. The challenge is to resolve the competing objectives of product quality and dye waste treatability by developing dyes which are more treatable, less dispersable, less persistent, and less offensive.

New Generation Dyestuffs. As indicated above, the textile industry needs a new generation of dyestuffs based on better treatability and higher exhaustion. These dyes would leave less color residue in wastewater and produce safer intermediates (see Fig. 1 and Fig. 2) while maintaining desirable tinctorial, cost, and fastness properties. The application properties of these dyes of the future should be adequate in terms of repeatability, compatibility with existing equipment, controllability, etc. This new generation of dyes should require fewer chemical dyeing assistants, especially salt, retarders, and accelerants. Fiber reactive dyes for cotton, which are the fastest growing class of dyes over the last 20 years, are promising for this major opportunity to reduce or eliminate some of the most intractable emerging problems enumerated above. These problems include:

> Color residues in dyeing, printing, and washing off wastewater
> Massive electrolyte discharges
> Aquatic toxicity
> Toxic air emissions
> Improving treatability of wastes
> Eliminating low levels of metals from wastewater

One approach which has been proposed is to make dyes more reactive. However, without higher affinity properties, the above goals can not be accomplished.

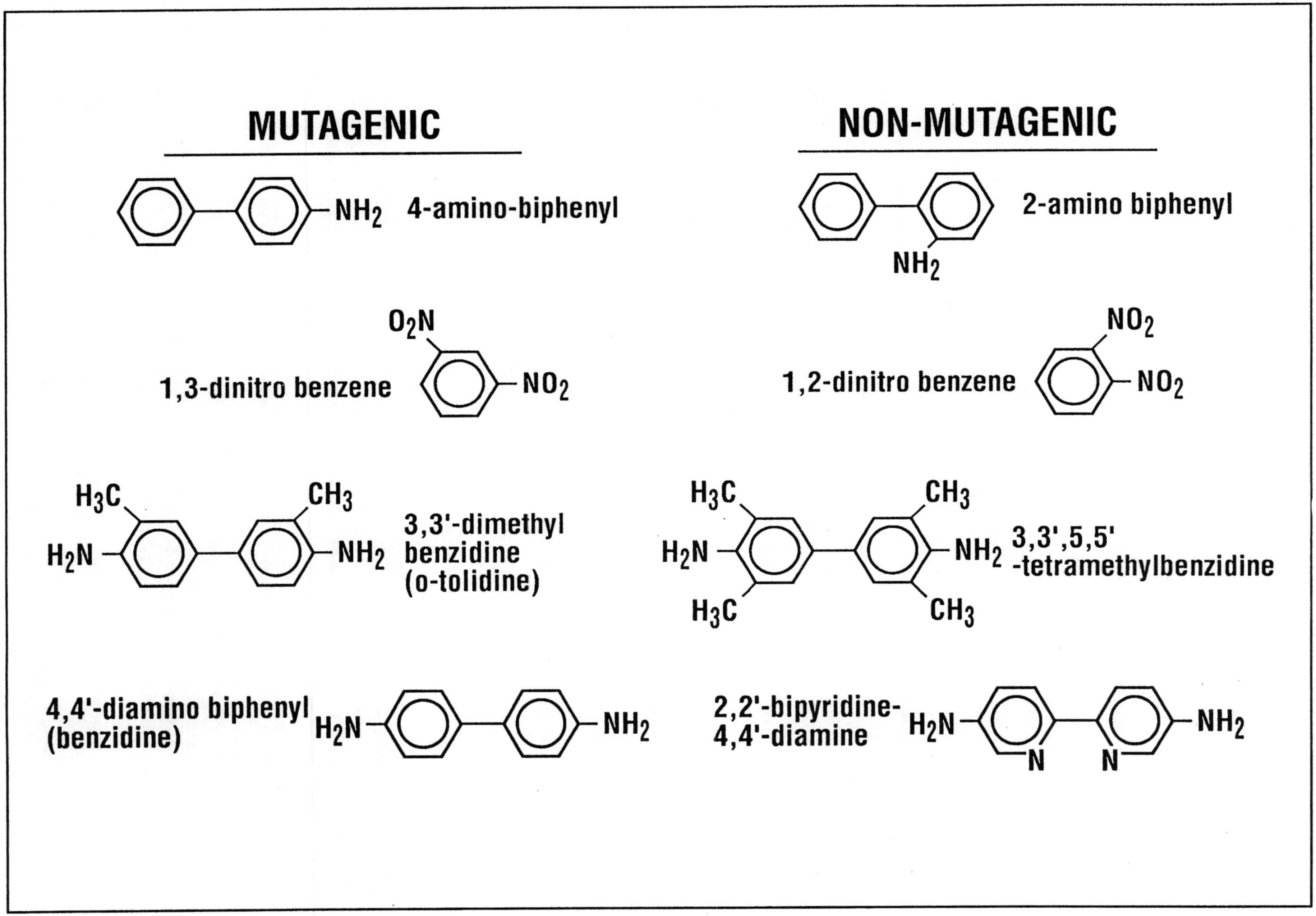

Figure 1. Safer Dye Intermediates

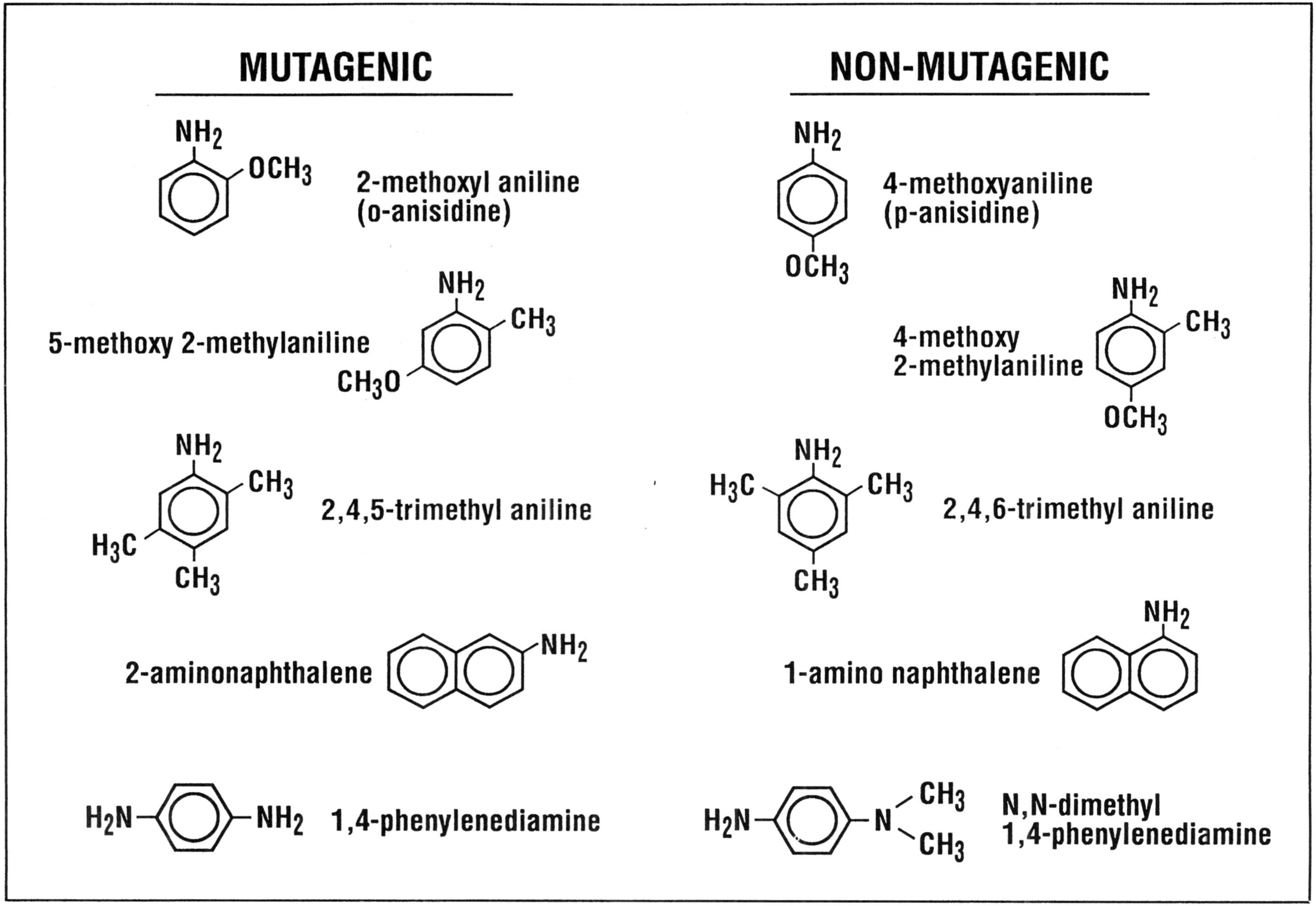

Figure 2. Safer Dye Intermediates

Another important opportunity is the development of azo dyes based on iron instead of other more harmful metal ions such as cobalt, nickel, chromium, lead, and zinc. This line of research has already shown promise in substituting iron for cobalt in Acid Red 182 and Acid Blue 171, and also iron for chromium in Acid Black 172 [19]. These new dyes are non-mutagenic and, of course, do not introduce undesirable metals in dyeing wastewater.

New Application Methods. There is also a need to develop new application methods especially for existing fiber-reactive dyes on cotton. One well known successful approach is pad-batch dyeing. Other methods are also needed, especially for tubular knits and other substrates not adaptable to pad-batch dyeing. Using alternative electrolytes or other methods to control dye affinity and exhaustion via solubility control and fiber zeta potential is another opportunity which merits investigation [20-23].

Proprietary Issues. In Section 2.2.3, the need for better disclosure of chemical constitution of chemical specialties is discussed. The same challenge exists for dyes. In the past, dyes have been classified by chemical structure and applications factors in the *Colour Index*. For nontechnical business reasons such as the withdrawal of support by many of the major dyestuff companies, this indexing system seems to be coming to an end. In the future, dyes will become more proprietary, like specialty products, with associated loss of information for the user. Lack of information in the future will make evaluation of pollution problems, substitutions, etc. even more difficult for textile mills because they will know less about the chemical constitution and structure of the dyes which they use.

2.2.3. Chemical Specialties.
Specialty textile processing assistants are a unique group of products, which are not understood very well by those outside the textile industry. These assistants are used in large quantities with relatively little information about their pollution potential. A recent listing [24] contained over 5,000 products in 100 categories which were marketed by 175 companies under 1,800 trade names. Each product is a proprietary blend of chemical commodities. The composition is not revealed to the user, nor are the pollution characteristics (e.g., aquatic toxicity).

Clearly, there is a need for disclosure of user information, which is, in a sense, a business issue, including potential incompatibilities with upstream process residues. However, there are also technical needs for better data on these mixtures, as well as more accurate aquatic toxicity data. Toxicity reduction programs are often frustrated because comparative evaluation of substitute chemicals is almost impossible. Test results for aquatic toxicity often correlate poorly among labs due to nonstandardized test conditions, species variations, and other difficulties.

The incentive to establish good pollution prevention programs is often economic. These programs are often justified through a "pollution prevention pays" type of thinking. The need for better risk/benefit assessment and more realistic waste goals, as well as realistic forecasting of benefits and liabilities is therefore critical. The challenge is to eliminate the barrier of poor technical information as described above. This will also contribute to the better technical understanding of processes discussed earlier.

2.2.4. Chemical Commodities.
In addition to specialties, the textile industry uses massive amounts of commodity chemicals, (e.g., acid, alkali, salt, warp size, fiber, and water). A typical cotton production facility might use a mass of commodities that is greater that the mass of product produced. It is not unusual to find cotton dyehouses which discharge 3,000 ppm of salt in the dye effluent wastewater. There are two needs in this area. The first challenge is to reduce the amounts of commodities, especially salt, required for dyeing. The second is to determine the trace impurities in commodities and to seek better sources of commodities, or better methods of manufacturing commodities, which reduce or eliminate offensive trace materials. The first need is discussed in Section 2.2.7, "Wet Processing."

The second major need, to identify and eliminate trace impurities from commodity raw materials, is crucial. Tests of textile wastewater [8] have clearly shown the presence of toxic materials which also have been detected as impurities in fibers and chemical commodities. Further work [18] shows that these toxic wastewater pollutants are present in significant amounts in high-volume raw materials (e.g., fibers), as well as in salt and alkalis. Major needs exist in this area, especially related to zinc in salt, low-level impurities in fibers (i.e., monomer, catalyst, delusterants), metals in caustic, and impurities (e.g., metals, organics) in raw water supplies.

The current practice of reusing waste commodities internally or from other industries can cut consumption and associated pollution discharge of commodities. This is generally a good practice, but impurities must also be considered. Caution and more information with respect to the above is needed in the reuse of commodities.

2.2.5. Yarn Formation: Spinning. In yarn spinning, routine waste production is minimal and virtually unavoidable. Also, most waste is recycled into further uses. Therefore, the main opportunities are related to additives and how they affect downstream processing, much like spin finishes discussed previously in Section 2.2.1. Targets for consideration include tint oversprays, winding emulsions, coning oils, lubricants, and, in some cases, biocides used for mildew suppression. This is an area where global attitudes can contribute greatly. In some cases, the yarn mill also applies sizes (e.g., PVA, starch, CMC, PVac) or lubricants, which will be reviewed under fabric formation in the next section. The need is to assure compatibility of additives with downstream processes, and to eliminate potential interferences to later processes. The challenge is to develop specific products for spinning mill use which have minimal downstream impact. This has been done to an extent with low BOD winding emulsions and waxes which do not pollute when removed, but not with other spinning additives noted above.

2.2.6. Fabric Formation. *Sizes and Lubricants* - Sizes and knitting oils added to yarns before and during fabric processing are one of the greatest routing intentionally created pollution stream in textiles [25]. Typically 6 percent or more of the weight of the goods is added as size or knitting oil, only to be removed and discarded in the next step of the process (i.e., preparation/desizing). Although size recovery is possible, knitting oils are never recovered. The amount of size material alone used in the USA is about 200 million pounds per year, making this one of the largest industrial waste materials. Although there are a few spectacularly successful size recovery systems in operation, the textile industry, for several valid reasons, makes limited use of size recovery. This is equivalent to thousands upon thousands of tons of intentionally created waste, making it along with water, salt, and cutting room waste the highest volume waste materials in textile manufacturing, and perhaps all U.S. manufacturing. The need is to dramatically reduce this waste. The challenges are to:

>Remove logistical and technical barriers to recycling and reclamation of sizes
Provide more incentives for recovery
Develop fabric forming machines and processes which require minimal amounts of sizes and knitting lubricants
Design yarns and fabric structures which require less sizes and lubricants to create

Recovery can only be accomplished with certain types of sizes, notably polyvinyl alcohol (PVA), which is roughly one third of the total size used. The remainder comprises non-recoverable sizes, such as starch. Less than one third of all PVA is recovered. There are several technical and business barriers, including the practice of applying PVA in mixtures with sizes that inhibit recovery, the high expense of shipping recovered PVA concentrate solutions, mixing of goods containing different sizes at the desizing plant, and a lack of understanding of recovery potential.

Regardless of the approach taken, a clear need exists to reduce what are undoubtedly two of the largest industrial waste streams in all of U.S. industrial manufacturing (i.e., warp sizes and knitting oils).

Fabric Structures. Certain indoor air quality factors are also a function of fabric design, structure, and formation. Preliminary modeling of pollutant exchange zones shows that fabric structure, air permeability, and velocity slip factors are all important parameters in the emission, sorption, and release of indoor air pollutants by textiles and multilayer textile containing products [26]. The opportunity is to design textile-containing products which inherently improve indoor air quality and minimize pollutant exchange by understanding fabric/air interactions at the microscopic level. Of course, another opportunity in indoor air quality is to design and produce fabrics which inherently require less chemical stabilization, and, thus, to eliminate the need for chemical finishing and achieve associated reductions in manufacturing pollution as well as lower risk of air emissions from applied chemicals [25]. This fabric design issue is discussed above in Section 2.1.3, Engineering Practices.

2.2.7. Wet Processing: Preparation, Dyeing, and Finishing.

Typically, over 50 percent of pollution from textile preparation, dyeing, and finishing processes results from removal of upstream processing residues which, if not removed, interfere with dyeing and finishing. A significant portion of pollution also results from application of chemical fabric stabilizers, stiffeners, softeners, etc. to adjust the characteristics of the fabric to suit the intended end use. Thus, pollution prevention in wet processing is intimately related to the global views advocated throughout this document. If it were not for the need to remove contaminants via preparation and to overcome technical design deficiencies via finishing, coloration (i.e., dyeing/printing) would be the main task in textile wet processing.

Pollution reduction has been utilized very successfully in wet processing [15]. Even so, there are still significant opportunities for advancement. Many of these opportunities relate to chemical specialties and commodities discussed previously, as well as education of smaller mills in the methods already successfully used by more sophisticated operations.

Preparation. The purpose of the preparation step is to remove contaminants which interfere with dyeing and finishing. Other than previously noted in Sections 2.1 and 2.2.3, the needs in this area are surprisingly few. In fact, the ultimate opportunity in fabric preparation is complete elimination of the process and, with it, approximately one half of all pollution from textile wet processing, not to mention a dramatic reduction in water use.

One rather surprising situation is the lack of heat-recovery systems in many textile operations. Quite a lot of hot water is discharged from preparation and dyeing operations, and many do not utilize heat recovery. This is a significant opportunity for energy recycling and thermal pollution reduction. The main barrier seems to be the perception of a lack of incentive to pursue heat recovery.

Dyeing. *Control* - One important need is to improve process control, especially in dyeing operations. The resulting color consistency, coupled with appropriate numerical color specifications, could provide the opportunity to cut adjacent garment (product) panels or parts from widely separated areas of fabric. This measure would diminish waste potential in cutting and sewing by improving marker efficiency. In order to do that, more uniform fabrics are needed. The use of controllable factors to offset uncontrollable variations and thus produce more consistent color repeats has been proven in the laboratory, using non-parametric methods such as neural network and fuzzy logic based real-time multi-channel adaptive control algorithms [27]. The economic and pollution control benefits of achieving this in commercial dyeing operations will be immense. This will not only apply to improved material utilization for piece dyed goods, but also yarn waste will be reduced in weaving and knitting through better yarn utilization in yarn dyed fabrics.

Models - Another need is better parametric models of complex dye systems, (e.g., fiber reactives) for control purposes. Quite sophisticated parametric thermodynamic and kinetic dyeing models are available for many dye classes [28]. However, there are still major opportunities to improve these models and to utilize them in parametric control algorithms or in "training" non-parametric control models. The challenge is to develop methods of parameter estimation which are simple and economical enough to apply in commerce. Another challenge is to develop parametric models which are simple enough to be useful in commerce, but, at the same time, robust and sophisticated enough to achieve highly accurate predictions of dyeing behavior. This may seem to be impossible, but recent work [28] indicates the possibility is real. A barrier to progress in this area is the perception that this research is too fundamental for industry to support, and, simultaneously, too applied to attract traditional basic research support.

Salt - One need which stands out in the near future for cotton dyeing is salt reduction. Currently, the salt requirements for fiber reactive dyes, which are the most important dye class for cotton, are 50 percent to 100 percent on weight of goods. It is not unusual to find textile mill effluents with 3,000 ppm salt from cotton dyeing operations. The total quantity of salt discharged from textile dyeing operations may be on the order of magnitude of 400 million pounds annually. It all becomes waste. The role of salt in dyeing is to promote dye exhaustion from the dye bath onto the fiber by decreasing the solubility of dyes in water, and by electrical effects including fiber zeta potential.(20-23) Reduction of salt in cotton dyeing processes usually results in lower dyebath exhaustion and, therefore, more color in dye wastewater. Reduction of salt from the current levels of up to 3000 ppm to desired limits of only 250 or less ppm will require significant developments in several areas including machinery, dyestuffs, and dye application processes.

Machine Cleaning - Currently, the textile industry schedules dyeing production based primarily on delivery times and cost factors. Two major pollution sources from continuous as well as batch operations are dumping unused portions of mixes and machine cleaning which may be necessary between shades. Machine cleaners are generally among the most toxic and offensive chemicals used in textile wet processing.

Dye machine cleaning requirements depend heavily on the sequencing of colors. Ideally, grouping colors within chroma families (e.g., red, yellow, blue), and sequencing from light to dark and from brighter/brilliant to duller/grayish. At present, "smart" scheduling systems which can minimize machine cleaning are not used for dyehouse scheduling. The need is to schedule dyeing production in such a way as to reduce pollution by minimizing machine cleaning as well as mix dumps. The opportunity is fairly straightforward, and the technical barriers to this are minimal. The barriers are discussed in Section 2.1, "General Needs" under the subheading, "Accurate Information."

Scheduling improvements is not the only way to reduce machine cleaning requirements. There are opportunities also to understand fouling and cleaning processes better and to develop:

> Dyeing systems which do not foul machines
> Machine configurations and surfaces (e.g., Teflon®) which are easier to clean without toxic chemicals
> Less toxic and more biodegradable machine cleaners

Robust Dyeing Systems - Poor dye work and associated off quality, rework, and pollution are often caused by the presence of upstream processing residues in fabric. The purpose of preparation is to remove these, as noted in Section 2.2.7. However, preparation processes sometimes are not completely successful in removing all contaminants. There is a need for dyeing systems which are more robust toward previously added materials (e.g., spin finishes, agricultural chemicals, sizes, oils, tints, and winding emulsions). Such systems could reduce or eliminate the need for preparation. The challenge is somehow to overcome the proprietary nature of specialties and globally select compatible processing assistants. The barriers to this are great, but the potential rewards of such an approach would be immense. There is a limited amount of work being conducted in this area by a consortium of textile companies and the North Carolina Division of Environmental Management.

Automation - Equipment automation has been major focus of textile process improvement over the last 10 years. At a recent International Textile Machinery Association exhibition, fewer than 100 companies showed dyeing machinery, but more than 150 showed microprocessor controllers, chemical dispensing systems, etc. Automation can produce good results in quality, productivity and pollution reduction, because routine waste levels are decreased, cleanup is easier, mixes are made more accurately, and human errors are reduced. On the other hand, the relative importance of malfunctions and spills increases. Also, there is a tendency for technical supervision to lose contact with automated processes. When a process is automated, routine maintenance becomes relatively more important. Maintenance and supervisory practices which have been used in the past with less automated systems may not be optimum when automation is installed. The challenge is to determine the optimum technology of pollution control for automated processes, and to determine how that differs from current practices.

Finishing. *Fabric Design* - Finishes are applied to provide desirable end use characteristics and to facilitate product formation (e.g., cutting and sewing). Proper engineering-oriented fabric designs can eliminate some or all of the need for finishing, particularly in terms of shrinkage, curling, and sewing lubricants. Also, it is possible to stabilize properly designed fabrics without chemicals by using mechanical finishes. Much recent finishing research has focused on chemical finishing, not mechanical. The opportunity is to substitute mechanical treatments (e.g., compacting, Sanforizing®) for chemical treatments. For these to be successful, it is necessary to correlate three items: fabrics designs which require less chemical stabilization, finishing machinery which can accomplish better end use performance, and compatible fabric specifications which accommodate the use of mechanical finishing through proper design of textile assemblies (e.g., garment constructions).

Indoor Air Quality - Finishing directly impacts indoor air quality because many finish chemicals contain low molecular weight, reactive materials (e.g., formaldehyde) which may later be emitted in the consumer's use area. Also, certain finishes (e.g., soil release, water repellent) change the fiber's critical surface energy and, thus, alter the sorption/reemission characteristics of fabrics. There is a potential opportunity to improve indoor air quality by understanding these factors [25].

2.2.8. Cut, Sew, and Fabrication.

Waste Utilization - The effects of design, planning, information and communication are clearly manifested in cutting room waste. Denim is an interesting example. About 800 million yards of denim are produced in the U.S. each year, at an average weight of approximately 12 ounces per linear yard, or a total of well over one half billion pounds. Fabric utilization efficiency in cutting and sewing ranges from the about 72 percent to 94 percent. The efficiency for cutting denim in particular is typically 84 percent or less. Cutting waste, therefore, represents about 16 percent of denim production or roughly 100 million pounds annually in the U.S. for denim alone.

Cutting pattern efficiency depends heavily on garment design factors such as shape and seam location; size assortment as required by retail sales; fabric width and other technical considerations. Cutting practices which minimize waste in view of these factors are well known, and material utilization departments are adept at optimizing cutting efficiency by relocating seams in garments, etc. Pollution prevention in this area is a highly developed science, using sophisticated computer algorithms and other techniques which, incidentally, may be very informative to other industries.

It is doubtful that commercially viable systems can be developed which will surpass current efforts at cutting waste minimization. However, advances could be made in the reutilization of the waste materials. Currently, denim-cutting waste is recycled into end uses such as paper making. If it were possible to develop manufacturing procedures to reclaim fiber from denim-cutting waste, not only could the cost of the raw fiber be saved, but dyeing would also be eliminated because indigo denim color would already be present in the reclaimed fiber. In addition to cutting room waste, post-consumer recycling of discarded blue jeans or other denim products could be combined to utilize the same technology. Another attractive feature of this opportunity is that the needs are entirely technical and there are minimal political or business barriers or entanglements with which to deal.

Indoor Air Quality - Most textiles are combined with other items in the final consumer product, and the combinations are essentially innumerable. Textile manufacturers generally do not know which components will be combined nor in what manner. For example, an upholstery fabric could be combined with other fabrics, batting, fiberfill, open or closed cell foams, or stiffening innerliners, etc. On the other hand, the product fabricator generally does not have good information about incompatible combinations in terms of emissions and sorption/re-emissions. This makes product design difficult design. Better information on combinations and synergisms will enhance indoor air quality.

2.2.9. Consumer Issues. The final link in the production chain is the consumer. Opportunities for source reduction include development of post-consumer recycling of textile products as mentioned above for denim, better installation and maintenance techniques to improve life expectancy of textile products, installation and information use for improved indoor air quality, and products which do not soil or do not show soil and, thus, require less cleaning solvents and aftermarket care requirements [25]. Discarded carpets are another potential source for post-consumer recycled fiber.

Better information for consumers about environmental impacts would require standardization of tests/terms such as "biodegradable." This better consumer information would make use more efficient and offset disinformation which, in some cases, now exists.

3. Business Opportunities and Needs for Pollution Prevention

In addition to technical needs and opportunities reviewed in Section 2, some business issues also deserve comment. Much pollution prevention success has been achieved *within* individual production units in textile operations. Even greater opportunities exist in pollution reduction programs which transcend production facility boundaries [6]. In many cases, the barriers and challenges are non-technical. These opportunities are reviewed here.

3.1. Priorities and Commitments. The need for global views and better information exchanges is controlled, to a large extent, by business priorities and commitments. The technical staff at a particular manufacturing site can develop and implement procedures for pollution reduction. But the need to develop global views of manufacturing can only be achieved by a higher level technical understanding across production unit boundaries. A prime requirement for this is better technical cognizance by those who operate across boundaries (i.e., management). Information exchange in this sense is not an end in itself, but only an enabling mechanism for actually understanding the predicament of other manufacturing stages. The opportunity is to develop special global business relationships among suppliers, various manufacturing sites, and customers to reduce pollution.

3.2. Marketing. Marketing of Wastes. In some cases, wastes are unavoidable so it is important to view waste as a by-product or secondary resource with potential value. Opportunities to market waste by-products should be sought. The business barrier is that the waste almost always sells for less profit than the primary products, therefore, the sales incentive is low. However, when costs of collection and disposal and potential liability are considered, the situation may be more profitable than it first seems. There is also a technical barrier in the sense that many operations are reluctant to buy waste materials as raw material inputs for quality and safety reasons. With so many disincentives, valuable opportunities may sometimes be overlooked based on generalized business views about marketing wastes.

Consumer Information. There is a need for more information on product use, installation, and combination synergisms from manufacturing to the customers. Marketing is a critical link in this chain. Some industries do an excellent job in this area. There is a need in textiles to emulate these other successful techniques (e.g., technical product bulletins, product specifications). As an example, a chair with upholstery will include particle board, foam, fiber fill, stiffeners, upholstery, paint, etc. all in combination. The fabricator of the chair does not know how combinations will interact because technical information bulletins are normally not available on textile fabrics. The manufacturers of each component usually have no information concerning component combinations.

3.3. Conflicting Goals. The conflicting technical goals of dye stability and dye waste treatability were reviewed in Section 2.2.2. There are even more difficult and hard to define conflicting goals in the non-technical arena [6]. Some of the more prevalent will be reviewed below.

Regulatory Barriers to Pollution Prevention. An often encountered example of non-technical conflicting goals which inhibit pollution prevention is the relationship of textiles and municipal sewer systems.

One particular dilemma, called the "water-conservation penalty," is illustrated by the following. In a simplified example, a mill discharges 20 million gallons per month of water with BOD of 400 ppm. The municipality charges $0.90 per hundred cubic feet (CCF) for the first 500 CCF, and $0.60 for all over 500 CCF. (Twenty million gallons is 26738 CCF.) The municipality also charges $0.50 per pound for BOD in excess of 250 ppm. The monthly sewer use charge computation for the stated situation, and for a 10 percent water conservation by the mill, are indicated by:

Item	As Stated	10 Percent Water Reduction
Water used	26738 CCF 20 million gallons 167 million lbs	24064 CCF 18 million gallons 150 million lbs
BOD	400 ppm 66720 pounds	444 ppm 66720 pounds
Water excess	26238 CCF	23564 CCF
BOD allowance	250 ppm	250 ppm
BOD excess	150 ppm 25020 pounds	194 ppm 29190 pounds
First 500 CCF Additional water Excess BOD	$450 $15,743 $12,510	$450 $14,138 $14,545
Total Sewer Charge	$28,703	$29,183

Although the mill in this example has reduced water use by 10 percent, the sewer use charge has increased. This is not an unusual or extreme example. In fact, it typifies many actual situations. Often, the results are even more extreme when other charges (e.g., COD, ammonia N, and TSS) are also included. The issue in this case is not technical. It is a simple matter of adopting a municipal sewer ordinance to encourage water conservation in textile operations.

A typical response by textile mills to the above is to adopt specialty processing assistants with lower BOD. Textile operations are often encouraged by such BOD surcharge regulations to make undesirable substitutions of non-degradable (low BOD) surfactants. These tend to pass through treatment systems and increase aquatic toxicity in treated effluent. Toxicity reduction in many cases is frustrated for several reasons:

> Evaluation of substitute chemicals is difficult because of poor correlation between labs
> Many chemical specialties are proprietary
> Technically correct substitutions are punished by administrative measures.

As discussed above, even dyes are now being converted to specialty status by the demise of the "Colour Index," a listing of generic dyestuffs.

Another example is waste segregation and capture. If there is no way to dispose of captured hazardous concentrated wastes (which is often the case), then the processor has little incentive to capture the waste for disposal in its concentrated form. Keeping hazardous waste out of sewers is often desirable but not rewarded. There is no legal way of disposing of hazardous waste in many states.

Situations such as the above are difficult to resolve with positive results, and attempts to do so usually develop into little more than long drawn out posturing. The need in this case is a greater understanding of the impact of regulations by those who write and adopt municipal sewer ordinances. The opportunity to accomplish genuine pollution prevention could be greatly advanced by genuine cooperation.

Quality Conflicts. Usually, the goals of economic gain and pollution prevention are very compatible, since high processing efficiency and low waste are essentially two sides of the same coin. Also, high-quality attitude of workers and pollution control through orderly work practices, etc. go hand in hand. However, occasionally these goals conflict. When this happens, the result can be one of the most difficult situations in which to implement a pollution prevention program.

Typically, this happens in very high-priced, high-quality, low-volume specialty manufacturing situations such as papermaking felts, coated fabrics, offset printing blankets, and high-quality printing. In these cases, the cost of waste is insignificant in terms of product value. Without economic incentives, progress is slow in pollution prevention. Also, the cost of product loss (i.e., off-quality, seconds) is so great that conversion efficiency is totally dominant and waste raw materials have essentially no value compared to product. The opportunity is to study these situations, and develop incentives and more applicable pollution prevention measures and techniques.

3.4. Risk Benefit Assessments. Better risk assessment and more realistic waste goals are needed. In many cases, one part of the risk/benefit balance is clear, but the other part is vague, nebulous, or poorly understood. Sometimes the barrier is poor technical understanding of processes. Another barrier is a cost system that views waste costs and liabilities as overhead items, not direct cost items [6].

3.5. Human Resources. Clearly, there is a need for more technical understanding among textile managers. Cost and liabilities (civil and criminal) are the responsibility of management in most cases, so it behooves the industry to develop informed management teams. Strangely, the largest textile universities have somehow interpreted this as a need to include more management in textile curricula by diminishing technical content of programs. The numbers of graduating textile management majors, who have minimal exposure to science, technology, and engineering, far surpasses the numbers of technical graduates, who have minimal exposure to business issues. The need is to bring educational criteria for various textile groups (i.e., management, design, engineering, chemistry, technology) closer together in terms of educational experiences. The challenge is to overcome the temptation to over-specialize at the undergraduate level.

In the same vein, there is a need to embody pollution prevention concepts in higher education. University education provides for interaction of engineers and chemists with managers and designers in general curricula to foster communications and a common perspective between these groups. The opportunity is to develop human resources to tackle tough future pollution reduction problems. The challenge is to create technical environmental competence in graduates. Few, if any, educational institutions have achieved this, but efforts are underway to implement such programs [16].

4. References

[1] Internet Gopher Database, North Carolina Textile Manufacturers' Association.

[2] EPA 440/1-74-022-a and EPA 440/1-79/022-b, Development Document for Effluent Limitations Guidelines and Standards for the Textile Mills, Superintendent of Documents, U.S. Government Printing Office, Washington, DC.

[3] B. Smith, Identification and Reduction of Pollution Sources in Textile Wet Processing (1987), and Pollution Prevention by Source Reduction in Textile Wet Processing (1988), North Carolina Division of Environmental Management, Raleigh, NC.

[4] E. Norman, Proceedings of the Conference on Pollution Prevention by Source Reduction in Textile Wet Processing, May 23-24, 1989, Raleigh, NC, Published by North Carolina Division of Environmental Management, Raleigh, NC.

[5] B. Smith and V. Bristow, Indoor Air Quality and Textiles: An Emerging Issue, American Dyestuff Reporter, Vol. 83, No. 1, 37 (1994).

[6] D. Chambers, Waste Minimization: A Corollary, Textile Chemist, and Colorist, Vol. 25, No. 9, p. 14 (1993).

[7] B. Smith, Dyeing and Printing Guide, America's Textiles International, p. 33, February, 1989.

[8] B. Smith, Identification and Reduction of Toxic Pollutants in Textile Wet Processing, North Carolina Division of Environmental Management, Raleigh, NC (1990).

[9] B. Smith, A Workbook for and Pollution Prevention by Source Reduction in Textile Wet Processing, North Carolina Division of Environmental Management, Raleigh, NC (1988).

[10] B. Smith, Pollutant Source Reduction Part I: Overview, American Dyestuff Reporter, Vol. 79, No. 3 (1989).

[11] B. Smith, Pollutant Source Reduction Part II: Chemical Handling, American Dyestuff Reporter, Vol. 79, No. 4 (1989).

[12] B. Smith, Pollutant Source Reduction Part IV: Process Alternatives, American Dyestuff Reporter, Vol. 79, No. 5 (1989).

[13] B. Smith, Pollutant Source Reduction Part IV: Audit Procedures, American Dyestuff Reporter, Vol. 79, No. 6 (1989).

[14] B. Smith, Source Reduction: An Alternative to Costly Waste Treatment, Americas Textiles International, March, 1992.

[15] G. Hunt, A Compendium of Pollution Prevention Case Histories, North Carolina Division of Environmental Management, Raleigh, NC.

[16] J. Lewis and J. Schurke, Incorporating Pollution Prevention Concepts into Higher Education, University of Washington Department of Ecology (1991).

[17] H. Kulube, Residual Components in Exhausted Textile Dyebaths, Master's Thesis, North Carolina State University College of Textiles, Raleigh, NC (1987).

[18] J. Lee, Unpublished Master's Thesis, In Preparation, North Carolina State University College of Textiles, Raleigh, NC.

[19] J. Sokolowska-Gajda *et al*, Textile Research Journal, In Press.

[20] S. Neale and A. Patel, Adsorption of Dyestuffs by Cellulose Part V: Effect of Various Electrolytes, Transactions of the Faraday Society, Vol. 30, 905 (1934).

[21] S. Iyer *et al*, Influence of Different Electrolytes on the Interaction of Chlorazol Sky Blue FF with Cotton, Textile Research Journal, Vol. 38, 693 (1968).

[22] S. Iyer and K. Subramanian, Influence of Electrolytes on the Absorption of Chlorazol Sky Blue FF on Viscose, Journal of the Society of Dyers and Colourists, Vol. 96, 185 (1980).

[23] H. Lokhande, Importance of Zeta-Potential in the Field of Dyeing of Textile Fabrics, Colourage Annual 1970, 11 (1970).

[24] A Buying Guide to Dyes, Pigments and Specialty Chemicals, Textile Chemist and Colorist, Vol. 25, No. 7 (1993).

[25] B. Smith, Reducing Pollution on Warp Sizing and Desizing, Textile Chemist and Colorist, Vol. 24, No. 6 (1992).

[26] B. Smith and V. Bristow, Indoor Air Quality and Textiles: An Emerging Issue, American Dyestuff Reporter, Vol. 83, No. 1, 37 (1994).

[27] B. Smith and J. Liu, Improving Computer Control of Batch Dyeing Operations, American Dyestuff Reporter, Vol. 82, No. 9, 17 (1993).

[28] R. McGregor, Ionizable Groupe in Fibers and Their Role in Dyeing, Textile Research Journal, Vol. 42, 536 (1972).

[30] B. Smith and J. Rucker, Water and Textile Wet Processing, American Dyestuff Reporter, Vol. 77, No. 7 (1987).

About the Author

Brent Smith is the Professor of Polymer and Textile Chemistry at North Carolina State University, College of Textiles.